GLUE IT!

Other TAB Books by the Authors

No. 1641   *Steel Homes*
No. 1681   *Ventilation: Your Secret Key to an Energy-Efficient Home*

To business and industry, these authors have written 23 books and 3000 magazine stories. We like to sell stories on special products for manufacturers. Contact us care of TAB BOOKS, Inc., P.O. Box 40, Blue Ridge Summit, PA 17214.

GLUE IT!

BY DRS. CARL & BARBARA GILES

TAB BOOKS Inc.

BLUE RIDGE SUMMIT, PA 17214

Liquid Nails is a trademark of the Macco Adhesives Group, Glidden Coatings & Resins
Division, SCM Corporation, 30400 Lakeland Blvd., Wickliffe, OH 44092.
"How to Restore Antique Furniture" reprinted with permision from *The Family Handyman* Magazine,
The Webb Co., 1999 Shepard Road, St. Paul, MN 55116. Photos by Gene Spatz.
"How to Pick and Use the Right Marine Adhesive" reprinted with permission from *Motor Boating & Sailing.*
"Which Glue Do You Use?" reprinted from *Fine Woodworking* magazine. © 1984 The Taunton Press, Inc.,
52 Church Hill Rd. Box 355, Newton, CT 06470.

FIRST EDITION

FIRST PRINTING

Library of Congress Cataloging in Publication Data

Giles, Carl H.
Glue it!

Includes index.
1. Glue. 2. Gluing. I. Giles, Barbara, 1944-
II. Title.
TP968.G55 1984   668'.3   84-8888
ISBN 0-8306-0201-1
ISBN 0-8306-1801-5 (pbk.)

Contents

Acknowledgments

We very much thank the following people for their assistance in writing this book:

Editor Gary Havens of *The Family Handyman Magazine* was most cooperative in furnishing photos and materials. Gene Spatz made the excellent photographs. The article is quoted in Chapter 3 with the permission of *The Family Handyman*, The Webb Co., 1999 Shepard Road, St. Paul, MN 55116.

Editor Peter A. Janssen of *Motor Boating & Sailing Magazine* graciously granted permission to use the article edited by Editor Bernard Gladstone on the marine use of adhesives.

Editor Jon Van De Water of *Industrial Equipment News* informed us of the importance of adhesives to industry.

George Mustoe graciously granted permission to use his fine two-part article on glues for woodworking which appeared in *Fine Woodworking Magazine*, The Taunton Press, Box 355, Newton, CT 06470. He also let us use excerpts from his original manuscript that gives excellent information on ensuring that glues will stick well. Dennis Danaher of the magazine was also most cooperative.

Helen Williams of Giles Communications, Inc. did a lot of research work on this book. Her telephone calls concerning it were very much appreciated.

The Pack Memorial Library in Asheville, North Carolina was most cooperative. So was the Black Mountain Branch Library. The reference department is especially applauded.

President Frank Noel of the Hermetite Products Corporation was cooperative in furnishing facts and photos concerning his excellent line of glues and adhesives.

President Edward A. Krause of Evans-St. Clair, Inc. contributed some valuable insights into the adhesive industry which are very much appreciated.

Introduction

Most of our lives are literally glued together. We will depend on glues and adhesives hundreds of times today. The chair you are sitting in right now is at least partially, if not completely, glued together.

Our furniture, many personal belongings, cars, and most houses and buildings are, to some degree, glued together. Our shoes are glued, as are purses, wallets, and millions of other items—including this book!

If all the glue on this planet suddenly came loose, the world would be in kit form. Just how much glue is there? One industrial statistic declares there are some 40 pounds of glue used each year for every person in America! Business and industry use hundreds of thousands of tons for an almost astronomical number of purposes. Technology and adhesion are stuck together very tightly. The world of industrial glues is vast and highly specialized compared to that of consumers.

The following pages offer information and guidelines on using glues and adhesives. Caulks and sealants are also in the field of fastening with chemicals and compounds; these are just adhesives or cements with additional properties for doing specialized jobs.

Very little has been written on glues and adhesives, considering their importance in the world. The magazines tend to repeat the same articles on gluing furniture, selecting glues for particular jobs, and the various craft uses. This is the first book that surveys glues and adhesives with facts for the consumer—sort of a consumer "glue bible," a first-aid manual for the home with information for practical applications.

Chapter 1

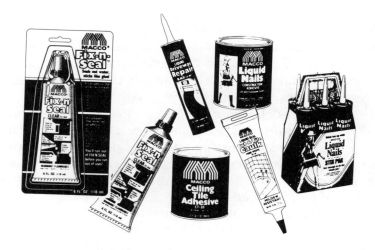

The World of Glues and Adhesives

Stickiness is as old as history. When Eve ate the apple, some juice probably tried to bond her fingers together. Prehistoric man was introduced to adhesive when he first sat down on some branches and found some sap had stuck to him.

That was good sticky stuff in the paint the Neanderthals used on the pictures they put on the cave walls in Western France tens of thousands of years ago. They are still very visible. (Some low-grade modern products would probably have peeled off in a few days.) Using hides, horns, plants, and an array of flora and fauna, man made various glues through the ages.

Ancient Egyptian tombs have yielded many items that utilized glues. Exquisite inlays of exotic woods and fine veneers still intact after eons attest to the uses of glues in the oldest civilizations. The earliest Greeks and Romans set the tiles in their temples with glues. Some marvelous mosaics have not lost a single tile from the original glue. Natives from every area of the world have used glues to make things for thousands of years. They have prepared their adhesives from various organic compounds. Their major source was the *collagen*, the protein of connective tissue, from fish and animal parts. Remote tribes in Africa and South America still make glues the same way their earliest ancestors did.

Furniture has been at least partially glued together from the beginning. Pieces in museums and private collections have been holding together for many centuries, attesting to the quality and durability of some of the first glues. The famed cabinetmakers—including Hepplewhite, Chippendale, Phyfe, and Sheraton—used glue in building their fine furniture.

Animal glues came into being when tribes found they could process hides, skins, bones, and sinew into sticky substances that would hold things together. These tend to be much thicker than those made from fish heads, bones, and skin. Early man discovered that he could make a fish glue—eventually named *isinglass* or *ichthcol*—from the inner skins of the swimming bladders of various fish. It is a white, tasteless gelatin with fairly good sticking qualities. Vegetable glues were prepared by ancient peoples from various plant matter. Almost every section of the world has some flora that seems to be sticky by nature, or they have the substances that may be converted to produce sticky preparations.

The first boats and ships were built thousands of years ago. The materials used to caulk seams and help bond boards and timbers together were pitches, rather than marine glues. Today there are many marine glues.

TRUE GLUES

Basically, there are only three true glues that do not contain chemicals, compounds, and high technology substances: hide or skin glue, bone glue, and fish glue. All are actually impure forms of gelatin. There are many other sticky products that are sometimes called glues. Technically, they are usually gums, cements, or adhesives. Glues and adhesives are so closely related on the consumer level today that the words are synonymous.

True glues come from breaking down protein matter in the tissues and bones of animals. Specifically, collagen is the protein of connective tissue. Heating these tissues in water results in part of their structure dissolving into a clear solution. When the solution cools into a solid, jelly-like mass, it is called gelatin. Although it resembles a food gelatin, it is not pure. When the gelatin dries, it will dissolve in water to form a liquid glue with strong adhesive capacities. It looks like a transparent mass.

True glue manufacturers get their raw materials from slaughterhouses, meat packers, and tanneries. Their raw materials may include scraps of hide, ears, tails, tendons, feet, and bones. Bones, scales, and skins from fish are also used in the fish glues. The same steps are used essentially regardless of whether the glues are made from bones, fish, or hides. The ingredients are prepared, the stock is heated, it is treated to remove impurities, and then the solution is dried.

Various methods are used to prepare glue. One of the most basic involves washing the hides first to remove all dirt, then they are soaked until they are soft. In successive baths in water, increasing amounts of lime are added to make the hides swell. Huge washing machines remove most of the lime, and the hides are ready for cooking. Cooking breaks down the protein matter and changes it into glue. This is a very precise process because if the hides are cooked too long or at too high a temperature, the quality of the glue will be destroyed. Depending on the objectives, the cooking may be done in an open tank or a pressure tank, the latter being widely used for untreated bones. Steam coils run through the large open tanks. The extracted liquid is drained and heated again until the glue becomes thick.

Clear glues are obtained by chemical or mechanical means. Alum may be added, or an acid following by egg albumin. These form a precipitate that removes the unwanted particles and leaves the glue clear. Some processors use mechanical filters, running the glue through paper or bone char, to make it clear.

Making glues from bones is somewhat more complex. After degreasing with solvents, an eight percent hydrochloric acid solution is applied to the bones to dissolve the mineral constitutents, primarily calcium phosphate. This leaves a soft, cartilaginous material—again, this is collagen—which retains the original shape of the bone fragment being treated. After the residual acid is washed from this soft collagen, it is dried, resulting in commercial ossein.

Herds of wild horses went to the glue factories in great numbers during the 1920s and '30s. The Marilyn Monroe and Clark Gable movie *The Misfits* graphically and romatically depicted how some people went into Nevada and other states and captured horses and sold them to the glue manufacturers.

CHEMICAL GLUES

Although fish and animal glues are still in use, they can not compete in holding power with the high-technology glues and adhesives of today, which number in the thousands. Most of them are for commercial and industrial uses. Amazingly, there are only some six types of glue available over the counter on a retail basis.

Contact Adhesives

"Contact adhesives are designed to make a fast bond between two materials upon contact," according to the Evans-St. Clair Corporation, one of the oldest and largest glue and adhesive manufacturers for industrial uses. "By definition, contact cements are adhesives which are applied to two substrates, dried, and mated under only enough pressure to result in good contact. The bond is immediate and sufficiently strong to hold pieces together without further clamping, pressing, or airing."

Originally designed for high-pressure laminates to wood and metal—such as countertops—the contact bond adhesives have now been developed to cover a wide range of usage in bonding leather, cloth, plastic, and rubber in the shoe industry. Contact bonds are also used extensively in interior upholstery and weatherstripping in automobile and furniture applications.

Evans-St. Clair makes a variety of contact adhesives. They have high strength and excellent heat, moisture, grease, and oil resistance. Contact adhesives are offered in either flammable or nonflammable solvents including a super-safe water-borne. These adhesives come in a wide range of strengths and working times, from the most forgiving to the most economical. The firm has them available ranging from 55-gallon drums to pints.

Most contact adhesives require certain minimum temperatures before they will adhere. A cold workroom or storage of the material at cold temperatures prior to use can cause failure. Moisture on or in the substrates will affect the bond. Materials stored in cold areas and then brought into areas of high humidity will cause condensation on the surface. Materials stored in wet areas or areas of high humidity will increase the moisture content of the more porous substrates.

Contact adhesives contain dispersed solid particles, which greatly add to their performance. Even non-settling adhesives tend to settle out slightly on storage and should be stirred back into suspension prior to use. Stir all adhesives before using them. Contact adhesives may lose a great deal of their strength when kept in high-temperature storage. The containers should be kept tightly closed when not being used.

Surface preparation is extremely important when working with glues and adhesives. To work, the adhesive must come in contact with the substrates. It cannot if they are covered with dirt, dust, sawdust, oil, grease, moisture, or anything else. The substrates and the work area should be kept clean. Both surfaces must be clean, and the adhesive should be applied uniformly over the entire surfaces. To adhere properly, the adhesive-coated surfaces must come into complete contact with each other.

Remember that the adhesive must be on the surfaces. If the surfaces are uneven, they should be leveled first or sufficient adhesive applied to act as its own leveling coat. This latter operation should only be done to even the normal texture of the surfaces, nothing more. No adhesive can make up for a warped or scored substrate.

If a surface is porous, it will absorb most of the adhesive when it is first applied. Allow this first coat to dry, and apply another coat. As with most of life, when in doubt, you should use more adhesive. One of the most common failures in gluing is a starved glue line. It should always be adequately coated.

"A contact adhesive is a solution and/or dispersion of compounded solids in a carrier, normally a combination of solvents," Evans-St. Clair Corporation points out. The solvents must be completely removed for the coating to give a stong permanent bond on contact with a similarly coated substrate. A very common cause of failure is incomplete drying. One may test the surfaces of both pieces that are to be glued together to be sure they are dry. Take a piece of Kraft paper and press it lightly on each of the coated surfaces. If the pieces are dry, there will be no adhesion between the paper and the coating.

Various factors determine the amount of time the surface coatings take to dry. The nature of the substrates, temperature, humidity, and the air circulation in the area where the gluing is being done control the rate of drying time. These circumstances may vary from hour to hour, day to day, or week to week. Drying is not always uniform either. Just because the coating is dry in one place does not mean it is dry all over. It is best to test in several spots for dryness.

You can govern the drying time by forcing it. Putting a heat lamp over the surfaces that need drying will speed up the drying time. This process also allows you to determine on a pretty regular basis how long it takes for various objects to dry. You can then regulate your bonding time.

Just as soon as the surfaces are completely dry, they should be combined immediately. Each adhesive has an open time, a certain number of minutes in which the pieces may remain uncombined without affecting greatly the bond produced. This open time is a safety factor in that it allows a certain number of minutes for any unexpected delays before putting the pieces together. If you feel that you have waited too long, the surfaces should be coated again, and you start over.

If everything has been done perfectly to this stage, the strength of the bond is lastly determined by the uniformity and amount of pressure used in combining the pieces. This pressure, depending on the glue job being done, might range from a huge industrial nip-roll or a simple hand roller. The strength of the pressure depends on the strength of the materials, and the pressure must be uniform.

TYPES OF ADHESIVES

Elastomeric adhesives are available in solvent and water-borne configurations. They are used extensively in the manufacturing of automotive vehicles of all types, in the boat, ship, and marine world, and in building and construction. In laminating things such as cabinet doors and in hobbies and crafts, adhesives find a multitude of uses. They are widely used in bonding carpeting and upholstery and in trim applications. There are special adhesives that bond fiberglass to metal in insulation applications.

Solvent-Borne Adhesives

Solvent-borne adhesives may be included

in four main categories. There are SBR (styrene butadiene rubber) neoprene, nitrile, and natural rubber. When inflammable solvent-borne adhesives are needed, they are made with non-burning chlorinated solvents. The major assets of solvent-borne adhesives include their fast drying time and water resistance. They also have the ability to blend with many resins and elastomers. They bond well to most surfaces. Their high-speed bonding makes them essential to many manufacturing processes.

SBR adhesives have a relatively low-cost base that gives good moisture resistance with an elastic bond for joining flexible materials. They are widely used in the insulation field. Some of them have a long tack for multipurpose use in duct liners, and they adhere many materials to metal and wood. In a nonflammable film, they are used to join insulation in duct work. Some are excellent for weatherstripping and as trim adhesives.

Neoprene adhesives are especially strong with strengths up to a maximum of 2,000 pounds per square inch when used in construction panels. They have applications in putting on automobile trim. They are used in vinyl mats, weatherstripping, construction curtain wall and structural panels, formica counter bonding, roof membrane adhesives, pit liner adhesives, and they also have general-purpose uses in manufacturing.

The non-staining, non-discoloring neoprenes are used in building seats for cars and planes, for upholstery, and even for landau tops. The general-purpose water and oil-resistant bonds are used in the boat industry for bonding rugs to decks. The bus, van, and mobile home industries use them to bond floor mats since some also have the asset of being nonflammable as well.

Curtain walls and building panels that have honeycomb cores use these type adhesives for their many assets. They may be applied in spray, brush, machine roller, or curtain coating viscosities. They are especially suitable for load-bearing building panels, and they are adhering from the Arctic to the Equator without being affected by extreme heat or cold. Black and white coating sealers are used in patching and waterproofing rubber products.

Nitrile adhesives are oil, grease, and gasoline-resistant. They are often used to bond vinyls and in paint rollers because of their solvent resistance and resistance to softening by vinyl films. Attaching vinyl trim on seats and upholstery is one of their main jobs. In addition to their bonding felts and foams, they are good general-purpose adhesives for patching vinyl materials.

Natural rubber adhesives have tremendous tack, with stickiness to surfaces of days to many years. They are important in artist cements, pressure-sensitive adhesives for labels, sanding papers, and tapes. They are also good repair and curing adhesives. Masking tapes use a lot of these adhesives. They are outstanding heat resistance. They have to take the tremendous heat produced by the friction of a sander, for instance.

Water-Borne Adhesives

Water-bornes fall into three main classes: SBR (Styrene Butadiene Rubber) acrylic, and neoprene. They are nonflammable, nonpolluting, and nontoxic. These adhesives are usually in latex.

The acrylics are employed as fixture adhesives. They bond bathroom fixtures to ceramic or any other household bonding, and work well on wood, metal, and other materials. Some types are good for formica bonding jobs. When dried, a good curing acrylic latex is pressure-sensitive. Scotch tape is an example.

Some cure in a strong heat and are moisture-resistant.

Millions of the headliners in vehicles are attached with a SBR latex adhesive. It has a very long pressure-sensitive tack for a one coating assembly and bonds to many materials. Notice how neatly the headliners in most cars are fixed. Modern cars rarely have headliners drooping, but this was extremely common in cars just a generation or so ago.

Another adhesive in this group is used in building and construction for bonding insulation to duct piping. It is stable when frozen and after it thaws. Many glues and adhesives are not free/thaw stable. They come loose when subjected to freezing and thawing.

Soft-seam adhesives have a quick and short tack. Foam fabricators and furniture manufacturers use them extensively. They bond strong, do not harden or embrittle at the glue line, and they are dimple-free. Similar high-quality adhesives are used to bond vinyl or fabric in the manufacturing of seats, benches, and padded decks, among many other things.

These latex or urethane-type foams may be laminated by coating both surfaces. The bonding is instantaneous, and the pieces may be handled immediately. They are superior for all types of foam boxing and splicing. The bond forms within 15 seconds, and it is as strong as it is going to get in less than four minutes.

Insulation adhesives are designed specifically to bond fibrous glass insulation to sheet metal in heating and air conditioning equipment. They also bond bead board, felt, cardboard, cork, sponge, and foam rubber to metal and other surfaces.

Chapter 2

The Most Common Uses of Glue

Everything that is breakable will, according to Murphy's Law. But the right glue may even put Murphy back together. Let's take a room-by-room inventory of the house. While we are doing this, we will also consider the contents of every drawer, shelf, and closet.

LIVING ROOM

What kind of furniture do you have? Is it wood, solid wood, veneered, a combination of these, or is it made of metal, glass, and plastic? Are there tears, holes, or puncturers in the upholstery materials? To make any repairs with glues and adhesives, we must determine the types of materials with which we must deal. They will dictate the type of glue or adhesive that must be used.

Do any of the chairs appear to be loose in any way? How about the table legs? Is anything split or splintered? Is any of the trim pulling away from where it should be? Is there a hole too large anywhere where a screw has been removed? It should be filled properly and the screw put back so it will seat properly wthout being noticeable.

How about the picture and mirror frames? Are they strong and secure? Have all the holes in the wall been properly filled where pictures have been at one time? Is there any wall damage that needs attention with a filler adhesive? How about the door hinges? Do they fit well, or do they need to be removed and replaced with a filler to make the screws fit better?

Do the windows need a sealant or caulking to make them airtight? One of the most common leaks is round the casings. Some repair may be needed to prevent the loss of energy.

Are all the lamps and lighting fixtures in good repair, or do they need some glue or adhesive? Check the items on the tables and shelves to be sure none are in need of glue to make them perfect again. Have you put any things in drawers that are in need of fixing? Glue may be able to do the job.

DINING ROOM

Do any of the china, porcelain, or ceramic pieces need repair? Is there a handle that needs to be put back? How about the glasswhere? The handles on some serving utensils tend to come loose, and they can be secured with glue quite quickly. Check the dining table and chairs for any looseness, squeak, or other problems. Inspect the rest of your dining room pieces, and make the needed repairs if they may be done with glues or adhesives.

KITCHEN

There are scores of items in your cabinets and storage areas. There may be a piece that has needed just one drop of glue for a long time. You are frustrated each time you try to use the item because it needs fixing. A small piece broken off an appliance may be readily replaced with the right glue.

If an appliance needs to be improved in appearance, you might consider applying one of the hundreds of decorative trims available. You may want to build an enclosure around something to enhance the decor or attractiveness of the kitchen. Building an enclosure around a refrigerator, for instance, takes little time with glue doing the bulk of the work for you. Cabinets that are plain and dull may be instantly changed to handsome, more decorative ones with some of the many panels that may be quickly applied with glue to the doors.

Some people note that they have never found a glue that will withstand the heat and water of a dishwasher. Well, there are some that will.

DEN OR FAMILY ROOM

This room takes a lot of action and use,

and there is probably at least one thing that needs gluing. Recreational equipment often needs glue repair, and some of the original glue may have been inferior or the item has been subjected to conditions that tore loose the bond. Pool table felt takes the right kind of adhesive so it will not be damaged or wrinkle. If there is a fireplace or other heating source in the room, see if anything in its immediate areas needs some gluing. Fireplace mantels take a lot of heat, and the right glue must be used to keep the materials bonded. Family room floors get a lot of traffic, and any loose floor covering or tiles that can be fixed with glue or adhesive should be checked.

BATHROOM

Tubs, tiles, showers, and various other items in the bathroom can be fixed with glue or adhesive when something needs repair. The glues for this area of the house must certainly be water and moisture-resistant. Many household accidents occur in bathrooms, and some may be prevented when every fixture and object are in proper working order. Sealing around tubs is especially important since the water and moisture erode the wall where it fits flush to the tub. If this isn't kept sealed, it can cause a rather major problem. Water gets behind the walls and may cause rotting. Many bathroom floors have to be replaced due to such damage.

Chips and scratches on porcelain and other surfaces may be repaired with the right adhesives. There are some adhesives that will seal the cracks in porcelain and other materials. Make sure the handles on brushes and other bath necessities are as they should be. Does the decorative top to that perfume bottle need to be repaired now? It would look much better fixed rather than laying off to the side, where it has been for too long.

BEDROOMS

Are all the drawers to dressers and chests fitting as they should? Is the bed solid with no loose parts? A little glue here and there might make the room much more comfortable and livable. Are all the drawer pulls secured well? Some adhesive might make the screws fit more snugly and get rid of that loose action. Surely, there is a shoe or purse in the closet that needs just one drop of glue to be like new again. That old wallet in the drawer just has one little problem: It needs a good drop of leather glue.

Now that you have checked your bedroom, you need to look over the guest room. Give it the same scrutiny you gave your room. Try every drawer, inspect the furniture from top to bottom, and note any item that needs a bit of glue.

STOREROOMS

Millions of storerooms have items that need some glue repair. This is the reason they are in the storeroom anyway—you didn't have time to fix something so you put it in there. Make sure that if you repair things in the storeroom that the temperature is warm enough for the glue that will be used.

TOYS

Repairing a toy may mean joy for a child and yourself. Fixing a broken toy can mean building on the image of being a highly respected and responsible parent. The child sees that you are concerned enough to take the time to make the toy fun again. Almost all toys can be mended with the right glue regardless of whether they are made of hard or soft plastic, metal, or wood. It is important to make sure that the glue is nontoxic! The age of the child must be taken into consideration. It is a good rule to always use a nontoxic glue, regardless

of the age of the child. Having glue among your household repair tools should be standard for toys also. A toy that can be mended in a minute may give a child hours of enjoyment, and the parent much peace of mind.

Even some of the bigger riding toys can be repaired with some of the plastic and vinyl glues. Molded plastic is one of the most popular toy materials. Seats and pieces of a riding toy that get a lot of wear and pressure may be fixed with the right glue. Applying some little bright trim to a toy often gives it a great new meaning to a child. Even a small colorful addition with an adhesive gives the child something to show to his friends, and it becomes a source of enjoyment to the boy or girl.

Depending on the toy and the type of glue that will be used, you may have to just press the parts together and hold for a few seconds. Other glues will require clamping, bracing, or rubber bands to secure the pieces while they bond. Some types of tape will secure something while it is bonding, but you must be sure that the tape does not leave a sticky residue that will have to be removed.

MENDING GLASS AND POTTERY

China, glass, earthenware, and stoneware may be repaired with glue, preferably an epoxy designed for that particular purpose. These give the strongest bonds; most will hold up under repeated washings in hot water. Some people claim they have never found a glue that will withstand a dishwasher, though.

The cellulose-based adhesives will harden faster, but they will not hold as well as the epoxies. Some gap-filling cements may take care of filling chips, and some are very good adhesives. If a piece has several fragments, work the puzzle of getting them together perfectly first.

Strong-bonding epoxy adhesives are excellent for replacing broken handles. Always glue the small parts to the bigger ones before graduating to the main job of attaching the handle to the item. Build up the filling area slowly and carefully on a handle area, smoothing the adhesive with wet fingers. Even soapy water on your fingers will work well in working on the surfaces. Scrape off any excess filler. Leave the filler line slightly below the glazed surface. In 15 to 24 hours, you can apply a colored adhesive that will match the color of the object or handle area. Paint pigments can be mixed with the small amount of adhesive that will be used to cover the repaired area. Epoxies come out shiny smooth when they dry, which helps match the color of the vessel.

Small chips on cups or mugs can be fixed using a little epoxy adhesive mixed with the proper paint pigments to match the color needed. Taping a repaired area is necessary to get the best bond. This is especially important when working a broken handle back into place. An epoxy cement and filler will be needed on larger cavities left by chips or damage. Putting a coat of regular epoxy into the cavity before using the epoxy cement and filler will make for a better repair also.

Fine vases and decorator pieces may be repaired with the best epoxies. If the piece is valuable or has great sentimental value, you might want to take the piece to a professional glass, pottery, or porcelain shop that does repairs. Their fingers are used to working with these things while yours are not.

Glass usually breaks into an explosion of small fragments. If you can gather all the pieces, work the puzzle of getting them together. After you get them where they should go, arrange them in their order of rebuilding. Lightly sand the pieces where they are to be joined with a silicon carbide paper. Cement the smaller pieces to the bigger fragments first.

The drying and curing process can be speeded up considerably by putting the heat from a hair dryer on the work.

One of the most important things to do first in working with gluing glass back together is to build a jig to hold the object. If it is a glass, footed or otherwise, it can be set in a glass holder, like those that secure wine glasses in a wood frame. You may have to tape the thing down to get it where you can work with it.

BONDING PLASTICS

Bonding is the best way to assemble or repair plastic. Bonding beats nails, screws, or welding because it distributes the load and strength over the entire length of the joint. Dispensing from a bottle of glue is far less involved than other fastening methods. Fashioning a jig to hold the plastic pieces being bonded is the only other equipment needed.

Hard, glasslike thermoplastics, such as acrylic, polycarbonate, polystyrene, and the cellulosics, may be bonded best by solvent cementing, a fastening method unique to plastics. This technique is based on the fact that these very hard and slick plastics soften on contact with strong solvents. Most people think of this as a melting or welding process. One of the most common methods is to assemble the joint and put on a few drops of solvent, usually methylene dichloride, either pure or in combination with other substances. The solvent can be put on the seam with a brush or squeeze bottle. Capillary action pulls the solvent into the joint, where it softens the abutting edges. These mate together, and there is a bond integral with the plastic when the solvent evaporates.

In another version of this bonding, only one edge of the abutting plastics is softened, but to a much greater depth, producing usually an even stronger joint. The best way to use this technique is to soak the edge of the piece in a pan holding a shallow layer of solvent. When the pieces are assembled, solvent from the soaked section softens the adjacent surface. One piece gets the action going for both of them, the edges melting and hardening into a strong bond.

This solvent cementing is generally ideal for many of the glass-like plastics found in the kitchen or bathroom, and it works well to join furniture made of them. Other solvent sensitive plastics—including PVC—may also be joined. The only requirement for this to work well is that the parts must fit perfectly together without any gaps. The fusion does not fill gaps. Square and sand the edges to be joined. In some cases, the joints must be watertight.

One way to test for a perfect fit is to put a few drops of water onto one face of the joint of the pieces that are going together. Hold the joint together and watch the water. If it spreads evenly down the joint, then the fit is fine. If the water puddles in some spots, the edges need more sanding. Keep testing with the drops of water after getting the pieces to fit as well as they can. This is particularly valid when working with clear plastics.

Acrylics may also be joined with the two-part acrylic cements; the joint does not have to fit so precisely here as it has a syrupy consistency and will fill small gaps. Beveling one of the edges to create a triangular crevice will make good trough for the glue. A polyethylene syringe, which can be bought in hardware or drug stores, is good for injecting the acrylic.

Other adhesives will repair an assortment of plastics. Items cast from thermosetting resins may be bonded with epoxies; contact glues will fasten polyurethane foams, while ordinary white glue will join polystyrene foams. If neatness is not required, there are adhesives that will work on the solvent-sensitive plastics. Model glues will work on polystyrene, and the general purpose glues will usually work on cellulosics. There are special adhesives for particular plastics.

There are more adhesives for joining plastics to other materials than there are for putting plastics together. This broader range of adhesives includes those which give a pliant bond of plastic to wood, metal, or other materials. This allows for the different expansion and contraction rates of dissimilar materials, due to temperature changes. These flexible adhesives do not contain fillers, and those made of silicone or synthetic rubber work well on joints of plastic to other materials. Viscous plastic cements and white glue will join plastics to many porous materials like wood. The super glue line locks plastics to a lot of materials.

ALL-PURPOSE CEMENT

There are many all-purpose cements, glues, and adhesives, but most of them are not as all-encompassing as their marketing would have you believe. Weldit, made by Hermetite Products Inc. (6500 Glenway Avenue, Cincinnati, Ohio 45211), is one of the finest in the world. It is dependable for a multitude of uses, including joining many dissimilar materials.

"Weldit cements anything to anything," the product says on the yellow and white card to which the tube is attached. The crystal-clear cement will repair plastic, crockery, wood, metal, tile, leather, glass, china, fabric, and other materials. It is not affected by gas, oil, or water, and it never becomes brittle, the firm declares. Weldit stops leaks, caulks, and waterproofs, and can be used to repair various types of shoes and boots.

Weldit is simple to use. The surfaces being repaired must be clean and dry. On nonporous surfaces—glass, china, or metal—the cement is applied to both surfaces then pressed together. On porous surfaces—wood and leather—apply two thin coats approximately five minutes apart and then press together after the second coating. The bond dries in about two hours, and the full cure takes 12 hours for maximum strength. Don't spill or place wet cement on painted surfaces. If you do have an accident or need to get some off your hands, use acetone or nail polish remover to clean.

This all-purpose cement is a good thing to keep in the house because it has hundreds of uses. It is also an excellent product to keep in the car because it can repair many things there, too.

Chapter 3

Glue and Furniture

Woods are composed of some 60 percent cellulose and 28 percent lignin. These constitute the woody and fibrous cell walls of plants and trees, including the cementing materials between them. The remaining material gives woods their individual properties, such as the deep richness of fine walnut, the decay resistance of redwood, and the color of mahogany. The beauty of the various grains is another matter.

Hardwoods include oak, walnut, maple, birch, and mahogany, which come from broad-leafed deciduous trees. These are more durable, last longer, and are more expensive. They are best for furniture and because they have better surfaces for fine finishing and appearance.

Even though hardwoods are stronger and prettier, they must be handled with care from the time the tree is cut until the wood becomes furniture. The quality of the wood depends on how it is seasoned and stored. Technology uses various kiln-drying processes, and a lot of air-drying is still done. Wood must be given time to reach the moisture content under the conditions in which it will be used. This transition is important. Even how the wood is stacked may influence its quality. The sticks inserted between the boards while they dry may leave marks on the wood that will never come out. Water may collect around these stacking sticks, and the pressure does the rest, leaving permanent marks on the lumber.

Softwoods come from cone-bearing or evergreen trees. Pine, cedar, fir, hemlock, redwood, spruce, cypress, and basswood are used extensively in construction and in furniture. They tend to have more knots—hard brown pockets filled with a sticky substance—than hardwoods. When working with these woods, the sticky areas should be cleaned with turpentine and sealed with shellac. Many knots get loose and will fall out. These should be glued first and then shellac-coated before finishing. Rough, split ends are common to softwoods. Although redwood will not weather, most pines need protection from the elements.

Furniture is made primarily from the outer portions of trees because they have the fewest knots. Lumber from this part of the tree is termed *clear*. Towards the center of the tree, the number of knots increases. Knots are nothing more than places where branches were cut or fell off. Animals make marks, and some trees have diseased sections where scars are left.

STICKING DRAWERS

A split drawer bottom is a common problem that can be corrected by removing the drawer and gluing a canvas strip across the split to reinforce it. If a piece is missing, cut a strip to fit as closely as possible and glue it in place, clamping the piece so it will bond well.

Drawer disasters must be completely taken apart. Clean away the old glue and join the parts back together with a good glue, clamping it to ensure a strong bond. A combination of weights and clamps may be needed to put pressure on the drawer while the glue drives.

TIGHTENING JOINTS

Loose joints are also common, and they can be fixed with glue quickly in many cases. First the structural elements must be pulled apart—legs, rails, rungs, or whatever taken from their sockets—and the old glue removed. Apply new glue according to the directions for the type being used. To secure the best bond, always clamp, and make sure the clamp doesn't

damage the piece.

Cracks and breaks around joints are no major problem if you approach them correctly. Sometimes only glue is needed, but if the place is bad, some nails may also be required. Round pegs are good for such repairs; these require drilling a couple of holes. Coat each dowel hole with glue and insert the rod. Join the pieces and clamp them. Wrapping the repaired area with wax paper before applying the clamp will keep the clamp from sticking to the piece being fixed. Sometimes the area can be clamped with stout rope or cord wound tightly around the repaired place. A tourniquet-type clamp with a cord may work well. Breaks in nonsupporting elements on tables or chairs may be fixed without doweling, just glue.

VENEERS

Loose or blistered veneer is an alarming affliction that is running rampant. Most veneers are brittle and delicate at best. What starts as an almost microscopic tear or loose piece may result in major damage very quickly. When veneer starts to come loose, it will split or break very easily. The thinness of the material is amazing, and extreme care must be used in making any repairs to it.

Blisters or bubbles in veneer can be flattened; place a damp cloth over the area, then slit through the cloth with a very sharp knife or razor blade, slicing into the bubble. Put some glue into the slit area and onto the underside of the loosened veneer, using a toothpick or other small applicator. Press the veneer back into place. If a clamp can't be used and masking tape will not work either, setting a heavy weight on the area will exert enough pressure to make the bond. Leave the weight at least 24 hours.

Laminated coffee table tops tend to warp and curl if the bottom of the top is not laminated as well. This can be prevented by gluing a piece of laminate to the bottom. If the piece is not visible, it does not have to match, and it will give the support needed to stop the table top from self-destructing.

DOORS

Loose door joints can disable a piece of furniture. If a door is coming apart, it must be fixed quickly to prevent more damage, such as marring the piece where the door strikes the surface. Dismantling the door is the only way to do a neat repair job. Using tools carefully, take the door apart. Work from the back if possible. Remove all the old glue, and clean the area well.

Once the door is in pieces and thoroughly cleaned, it is ready to be glued. Using one of the best wood glues, assemble the door, clamping it at all points where it comes together to get a good bond. Most house doors are not solid. They are just wood sheets glued together here and there with some blocks. Some of them are no more than cardboard inside!

When hinges do not work well on doors, remove them. The usual problem is that the screws are not tight enough. Take some toothpicks or matchsticks, dip them in glue, and insert them into the screw holes. When the screws are put back, they will have a good hold.

FURNITURE PROJECTS

There are virtually unlimited plans, materials, kits, and books about building furniture. Glues and adhesives are instrumental in the construction of almost every piece. Even men and women who have had no woodworking experience can build high-quality pieces with the products and information available.

Starting with an outdoor lawn or patio piece is recommended by many glue au-thorities. Many tables and chairs for use outside are based on variations of the same design. Fixed-top tables or chairs that convert easily to tables may be built without any prior woodworking skills. Glue and dowels are major items in the construction of this furniture. The slats are attached with screws and glue. Naturally, a high-grade waterproof glue is needed, and the strength of the pieces will be most impressive when the gluing is done well.

Furniture Kits

Furnishings for living areas may be built as easily as outdoor pieces. Plans for making any piece are readily available at many of the same places that sell wood for home projects. The library is a free source for a vast number of furniture and many other projects. Better libraries stock the newest offerings in books and magazines, giving you a tremendous selection of things from which to choose.

For many home items, you need only the basic frame. Various decorative materials can be applied to pieces to make them outstanding improvements to your home. With the time and the plans at your disposal, things can be added to every room in the house.

Kitchen improvements may include cabinets—either adding to them or improving existing ones. Tables and chairs can be made, along with shelves, stools, racks, containers, wine racks, and various other items. In other rooms of the house, stereo and entertainment centers, chairs, tables, cabinets, alcove units, storage boxes, and other furnishings and accessories will add to living areas. In bedrooms, even beds can be built—a very wide range of them. Night tables, wall units, and closet improvements can be made. Children's toys and furniture can be made. All these projects involve working with glue, and this book details how to work with various glues.

Heavily padded chairs and sofas get a lot of movement and wallowing-around in by people; dining room chairs get a lot of action. With these padded-seat types, the joints tend to loosen after a time. Gluing corner blocks to the joint areas will help reinforce them. The blocks should be hardwood, and they must fit snugly. A block that does not fit well will do no good. Clamp the block securely, and consider adding a screw or two.

Casters, glides, and levelers split a lot of furniture legs at the bottom. The inserts need to be checked regularly, and preventive measures are recommended to ensure that a fine piece of furniture receives no damage from the casters. New casters might solve the problem if they are worn. If they are in good shape, the hole where they go needs attention. Cleaning out the hole is the first thing to do. If the pin does not fit well, then a dowel should be glued into the hole, completely filling it so that a new hole may be drilled precisely to fit the caster pin.

When a leg is split, remove the caster. With surgical care, get glue into the splits, and bind them tightly with masking tape or a band clamp. An oil can or glue injector will force glue into tiny places. A lot of stress is put on a leg when the piece of furniture moves just a little bit. This action is what splits most legs. When a joint needs just a little tightening, some thread, string, or cloth may be enough to make it fit as it should. Wrap the end of the rung or piece with a nylon thread or string that has been soaked in glue. Check to see that the connection is going to be secure before applying the glue. Putting a strip of cloth on the end with some strips coming out at different points may be the solution to making a tight joint fit. Test before using the glue to secure it.

Thin wood strips can be used to make joints fit more strongly. These wedges should be glued to keep them in place. This method of tightening works pretty well with square legs and parts. When there is a space open enough, a leg or rung may not have to be removed. Never dismantle a piece unless it is absolutely necessary. Injecting glue with a glue injector into the loose joint may correct it. Toothpicks jammed into the gap will also help strengthen the fit. Fill the space with all the glue and wood that can be forced into the opening without it becoming visible.

One way to get glue into a joint without taking it apart requires some precision drilling. Put a 1/16 inch hole at an angle into the loose area. Then inject the glue through this tunnel. Tamping a small piece of wood into the hole (while not forcing out the glue) will often do a superior job of repairing a loose joint. Make sure the hole can be drilled so that it will not be visible when the piece of furniture is in its normal position.

There are a few woodworking techniques that might help in gluing furniture. Soaking the end pieces of wood that fit into joints in warm water for a few minutes and then letting them dry will open the pores in the wood. Drying on the top of a radiator is good, but drying in the sun is all right, too. Just warming the wood by one of these means will make it more porous to better accept the glue.

Removing old glue globs with warm vinegar is one of the best ways to clean pieces. When drawers do not work well, some wax, soap, or a silicone spray will make them glide much easier, reducing stress and friction.

Patching solid wood is not too challenging if you make a cardboard template of the exact size needed. The wood may be a low grade if it is not exposed, but if it is visible, the wood should match as much as possible. The area requiring the patch needs to be shaped to a particular pattern. A square, rectangle, triangle, or other shape should be made first. Then make the template to fit. Use the template to make the needed patch, cutting it out with a fine-toothed coping saw. Apply glue to the cavity and/or to the patch according to the directions. Put a weight or pressure on the piece until it bonds. Odd-shaped patches seem to blend into the surface of a piece and are not so visible.

Outdoor cushions can be re-covered using a good waterproof cement. Cushions can be made by using polyurethane foam, a leather-like vinyl, and a few tools. Put down the foam and cut it five inches larger all around than the piece. You can make neat corners by folding the excess up and over the foam. Cement the corner pieces where they make contact. After the recommended amount of time, press the substrates together. Tape will help hold the adhesive securely while it is curing. When the job is dry, cut a piece of vinyl to cover the back side, cementing it in place in the same manner.

If all else seems to fail in fixing a piece of furniture, you might try putting a dowel through the whole piece that has the problem. Drilling a hole right through the damaged area—or maybe even closely above or below it—and sticking a dowel completely through it until it is flush on each side might be the answer. Cut off the excess until it fits flush with the piece. Dipping the dowel in glue will give it what it needs. If you are using one of the one-drop glues, just follow directions.

Sometimes dowels break in furniture joints. The ends must be drilled out to the desired size of the replacement dowel. Generally, the new dowel should normally be the same size as the one that broke. Making it fit better and using an excellent glue will probably prevent the dowel from breaking again. When working with a piece of furniture that has been dismantled, it is best to put the smaller pieces back on the larger ones first. These can be curing while you work with other pieces. Allowing them to bond well before attaching them to the main part of the piece will make them much easier to handle, and they will not be moved out of position.

Repairing Turned Legs

Fancy legs on furniture that need repair involve a certain type of mending. If a curved leg on a table or other piece is broken, leaving a ragged break, a dowel joint should be used. First glue the break together, clamping the joint while it bonds. After it is well-cured, saw the leg apart below the original break. Tap a brad or small finishing nail into the lower section and push the two pieces together perfectly. Then pull them apart, and the brad or nail will have marked the spot for you where the dowel should be put. Drill holes in both sections on these markings to accept a dowel. Put glue in both holes, then insert the dowel to join the pieces. Let the glue set until the next day before putting any weight on the piece.

A split turned leg should be separated into two pieces. Apply glue to both sections and clamp them—tightly wound rope gives a good secure fit. After the glue has set for several hours, drill a hole from the bottom of the leg right on up past the repaired area. Get glue all the way down into the hole, and then insert the dowel after coating it with more glue. Don't just overflow the hole with glue, but make sure it is well-coated. After the glue has set for several hours, cut away the piece of dowel at the base of the leg flush.

ANTIQUES AND GLUE

Putting some glue on fine furniture is one thing. Mending antique furniture with it is an exacting dimension. With more and more people buying antiques instead of new furniture, it is essential that one put adhesives and antiques in perspective. If you purchase a new piece of furniture for $200, it may be worth half that the next day! It is like driving a new car around the block and having it depreciate several thousand dollars. Buy a $200 antique piece, and it may be worth *more* money the next day.

Many people work on their own antiques, especially if the work doesn't require special tools or special skills. The following article is one of the finest ever done on working with antiques. It appeared in *The Family Handyman* in January 1983. Editor-in-Chief Gary Havens of the magazine was most cooperative in providing this information.

Professionals at Sotheby's Restoration, a division of the famous New York City auction house, shared their years of experience with *The Family Handyman* readers. The article was written by Sherry Romeo, and the photographs were made by Gene Spatz.

HOW TO RESTORE ANTIQUE FURNITURE

A $100,000 Louis XIV table needs to be reglued. An antique set of fire-damaged chairs, worth more than most brand new cars, is about to be refinished. An antique mirror needs some inlay replaced. Sotheby's Restoration in New York City is like a hospital for wounded antiques too valuable to discard, too cared about to be left as they are. At a prestigious shop like Southeby's, the work must be more than professional: it must aim for perfection.

The Family Handyman interviewed Sotheby's restorers to learn how the pros solve some very common problems: Cracks in wood, loosened joints, warped boards, chipped gilding, delaminated veneer, or dulled finishes. Here's how they do it.

Regluing Chairs

John "Scottie" Scott has been a cabinet-maker and restorer for 46 years. He spent five years as an apprentice in his native country, Jamaica, learning his trade—how to boil glue, how to plane, and how to handle tools, not to mention sweeping and cleaning the shop. Then he immigrated to London, where he worked for 11 years at a very prestigious restoring firm before coming to the U.S. and going to work for Sotheby's Restoration division.

It should take about three hours to completely take apart and reglue a chair, estimates John "Scottie" Scott. The chairs used in these photos were fire-damaged, and it is worth noting that all cleaning and refinishing are done after the chair is reglued.

The Expert's Instructions

Step 1: Number all the pieces. All the pieces that fit together should have the same number—not different numbers. You may be tempted to skip this step, but it will prevent mistakes and make it much faster to reassemble. Hint: Use small pieces of masking tape; they're easily removed when the job is done Fig. 3-1).

Step 2: Use a mallet to completely disassemble the piece of furniture to be reglued (Fig. 3-2).

Step 3: Wash all the holes and places where there is old glue with a brush dipped in warm water. The idea is to soften the old glue so it will mix with the new glue and hold the wood better. There is no need to remove old glue, according to Scott. "Take your time at this stage, since it's very important to get water everywhere" (Fig. 3-3).

Step 4: Fit up the chair before regluing, then take it apart again.

Step 5: Apply glue in the following manner: First glue all the female parts; then glue the male parts. Now fit them together. Wipe off excess glue.

Step 6: Tighten the chair using rope or

Fig. 3-1. Don't rely on memory; number all parts before assembling, noting that the same number is used for all parts that connect. Courtesy *The Family Handyman.*

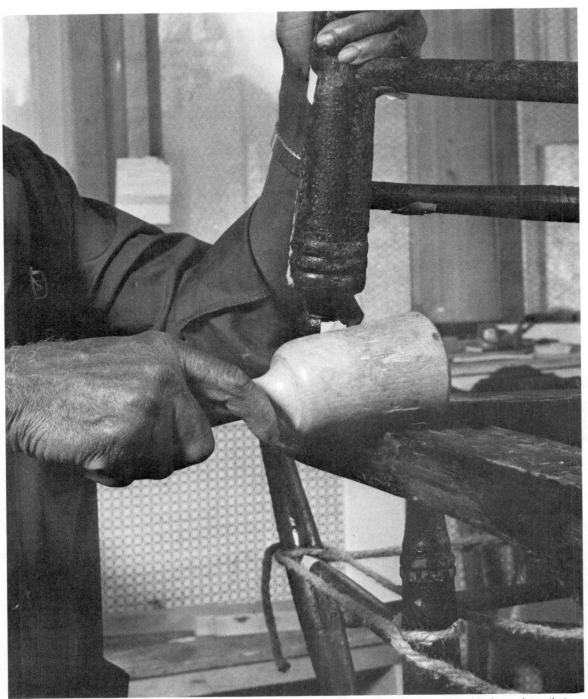

Fig. 3-2. Tap the chair apart with a mallet, not forcing apart any pieces that are firmly glued, only regluing those pieces that are loose. Courtesy *The Family Handyman*.

Fig. 3-3. Swab all holes and areas where there's old glue. You don't want to remove old glue, just soften it so it will mix with new glue. Courtesy *The Family Handyman*.

cord that is cut into lengths long enough to make a loop that goes around two legs. Repeat this procedure four times until all four legs are looped with the cord (Fig. 3-4).

Step 7: Cut four wooden dowels to a length longer than the middle of the chair pin. Now insert one dowel into the middle between each loop and twist until tight. Note: It will "lock" itself against the pin after tightening (Fig. 3-5).

Step 8: Stand the chair upright and place it on top of a board. Clamp the seat of the chair to the board.

Step 9: Double-check that the legs are square by measuring the diagonal of the legs. Make any necessary adjustments (Fig. 3-6).

Step 10: Let glue dry for four to five hours. "The glue may be dry at this point, but it's wiser to wait one day before actually using the chair," advises Scott.

Step 11: Do any cleaning, refinishing, and polishing at this time.

Gilding

Allison Armstrong was a sixth-grade schoolteacher when she decided to do something in the art field. She got a job in Washington, DC., and started learning restoration. "My apprenticeship was a very poor time of my life. Apprentices start at the minimum wage and have to do unimportant things for years before they get a chance to work on something important." Armstrong now works on expensive items, such as a $250,000 lacquered bureau/bookcase.

Only an expert should attempt gilding, but

Fig. 3-4. Cut lengths of cord or rope and make loops that are snugly knotted, but not too tight, repeating until all legs are looped. Courtesy *The Family Handyman*.

Fig. 3-5. Cut a dowel length that is longer than the length of a pin, insert it into the middle of the rope and twist until it's tight. Courtesy *The Family Handyman*.

Fig. 3-6. After the chair is clamped on a board, a check is made to be sure all the legs are square by measuring on diagonal. Courtesy *The Family Handyman*.

anyone can touch up gilding, according to Allison Armstrong, head of the decorative department, who oversees all of the shop's lacquering, gilding, and hand-painting. "This is a simple, straightforward procedure that will restore a time-worn or chipped piece to its original luster in less than an hour."

The Expert's Instructions

Step 1: Clean the area you'll be working on with a stiff brush dipped in turpentine. Wipe the area dry with a cheesecloth.

Step 2: Reglue any loose chips on the piece and fill in any voids where you can see the wood under the gliding. Use DAP or make up a mixture of five parts spackling compound to one part white glue (the glue makes the spackling harder). Hint: Armstrong mixes up a fresh compound in a disposable paper cup (Fig. 3-7).

Step 3: Let the work dry for approximately 30 minutes or until it is completely dry. The thicker the coat, the longer it will take.

Step 4: Sand the area smooth. This is done in two stages. Armstrong recommends 320-grit wet-and-dry sandpaper to start and finishes with 400-grit sandpaper for the extra smoothness required for professional results (Fig. 3-8).

Step 5: Paint everything with a one part shellac to two parts alcohol mixture (use yellow shellac, not clear, for best results). This will brighten up all of the guilding. Let it dry for about one minute before proceeding.

Step 6: Paint the gilding on the sandpapered area. To make the gliding, use bronze

Fig. 3-7. Clean the area to be worked on with turps and then wipe dry, filling exposed wood with a spackling mixture and smoothing it out. Courtesy *The Family Handyman*.

Fig. 3-8. Armstrong recommends 320-grit wet-and-dry sandpaper to begin, following with 400-grit to make the surface extra smooth. Courtesy *The Family Handyman*.

powder color mixed on a pallet with shellac and alcohol. You have to mix together several bronze powder colors to match the original gilding.

Step 7: Blend and tone the gilding with oil paints. "Eyeball it for the best results," advises Armstrong. "Then blend it into surrounding area with your finger." Let it dry for a few days or use a mixture of turpentine and cobalt dryer to speed up the process. Figure 3-9 shows the Armstrong touch.

Step 8 (optional): In a few days you may want to brighten up the gilding by again brushing everything with the shellac and alcohol mixture.

Regluing Veneer

Alan Hamilton attended Syracuse Univer-

sity as an industrial design major. At the end of his senior year he took a course in woodworking "to relieve the stress." He liked it so much he got a job in woodworking, and in 1981 he came to Sotheby's.

Squirting on some glue is simple; deciding which glue to use and getting what you want glued so it's precisely in place is not always so simple. Cabinetmaker Alan Hamilton is a keen advocate of familiarizing yourself with the technical information available, knowing the properties of the materials you are working with, and the applying a seat-of-the-pants approach to solving a gluing problem.

"There's an art to wading through the superfluous details that hover around a problem but don't solve it," says Alan, who points

Fig. 3-9. The trick to invisible retouching of gilding is to be able to mix bronze powder color as close to the original color as possible. Courtesy *The Family Handyman*.

out that in restoration, the basic problem is that you aren't the original gluer but are trying to reglue someone else's work that has failed. The task calls for good detective skills.

Another dilemma: Which glue to use? The restorers at Sotheby's spurn "modern" glues such as epoxies because they are insoluble. "You'd better get it in the right place the first time or you have a permanent problem, and, if at some future time there's a need to reglue, it may be difficult or impossible to do it without damaging or destroying the surrounding area," explains Hamilton.

"White glue, such as Elmer's, or hot glue is what we use," says the cabinetmaker. "Time is the factor that usually determines which glue you use. It takes white glue up to one and one-half hours to dry, which gives you the time

to work with something, such as a chair or table. On the other hand, there are times when you want almost instant bonding, and this is when you want to use hot glue, which dries in about five minutes."

Hamilton demonstrated how to reglue a loose piece of veneer. You can apply this technique to regluing any loose piece. Hint: Not all loose areas are obvious. To locate where the veneer has separated from the wood but not come off, take your knuckle and tap the wood. If it is still glued, there will be a solid sound. Loose areas will give off a hollow sound.

The Expert's Instructions

Step 1: Locate any loose areas. If they are not accessible, use a knife to split the wood

26

so you can insert the glue (Fig. 3-10).

Step 2: Apply a generous amount of glue to your tool, which should be small and flat enough to insert under the loose piece without tearing it up. Figure 3-11 shows the amount of glue being used.

Step 3: Slowly work the glue in until it has coated all of the loose area. For the best results, get the glue as far in as possible. Figure 3-12 shows this surgical operation.

Step 4: Squeeze out the excess glue from underneath and wipe up the excess with cotton, wool, or natural-fiber rag dipped in warm water (for hot glue, use even warmer water). The idea here is to remove the excess glue as soon as possible to prevent any problems when staining. Residual glue will cause light spots on wood.

Step 5: Clamp the wood down. A trick used by the Sotheby pros: Place a sheet of waxed paper over the wood surface before positioning a block of hardwood, and the clamp into place. The waxed paper keeps the block from sticking to the glued area and also prevents surface damage.

Cleaning and Polishing

Chief Finisher John 'Old John' Torretta has been in restoring since 1947, when he got out of the Army. "I had to do something, so I applied for this job in restoration. They asked me if I knew anything about furniture. I answered, 'Yeah, you sit on a chair.' I was hired—but I did all the dirt work and garbage."

Since then John has worked on one of Abraham Lincoln's beds, Napoleon's desk, and

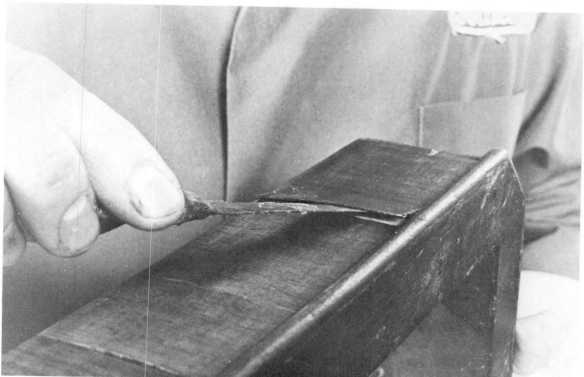

Fig. 3-10. Use a flat tool such as a pocketknife, nail file, brass shim stock, or a piece of a cut-up tin can to get under loose veneer surfaces. Courtesy *The Family Handyman.*

27

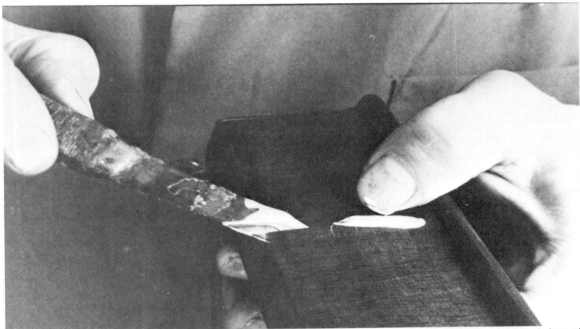

Fig. 3-11. The idea is to work fast and yet carefully when applying a generous amount of glue under the loosened piece of veneer. Courtesy *The Family Handyman*.

Fig. 3-12. Slip a knife under the loose veneer, wiping off any excess glue, clamping on a piece of wood and then letting dry. Courtesy *The Family Handyman*.

furniture owned by Madame DuBarry.

John's advice to aspiring restorers: "In this business you have to like it or get out. You're not working at the five-and-dime."

The idea is to clean furniture without causing any damage. "The problem is, it isn't all bingo," points out John. "You have to follow the rules up to a point and, after that, use common sense and the process of elimination."

"I don't follow the book; I threw it away a long time ago," shrugs Torretta, who prefers using turpentine to clean furniture. "Sure, they sell things to remove grime and wax, but I use turpentine because it doesn't harm anything and it can be wiped off with cheesecloth. You may have a white streak appear, which people worry about, but that is silly because it always disappears when you wax." Note: Turpentine is a highly flammable substance and should only be used in a well-ventilated area; never smoke around any volatile substance.

To watch Torretta clean and wax a simple chair, such as the one in these photographs, looks easy. "It is, it is," he sighs, "until what you are doing doesn't work. Let's say you are not sure whether you are cleaning off shellac or varnish. First you try alcohol. That will remove shellac. If that doesn't work, it may be varnish, so you should apply lacquer thinner. And sometimes, no matter how much you know, it's a mess and you plain have to give up and strip the whole thing."

The Expert's Instructions

Step 1: Wipe off the piece before you start only if it's dusty.

Step 2: Apply turpentine with a clean rag (cheesecloth is good, but an old sheet will work too; avoid synthetic or nonabsorbent materials). Wipe off well. "The idea is to remove most of the dirt, grime, oil and wax. If the piece you are working on is already in good shape, just go ahead and wax it."

Step 2A (optional): If the finish is not in good shape, now is the time to touch it up. Also, if you have a stain that doesn't come out with turpentine, put some alcohol on a cotton swab and carefully work the stain off. "Don't use too much alcohol, or it will run and stain," Torretta warns.

Step 3: Apply paste wax with a good stiff brush. You don't have to wait for the wax to dry to start going over it with a dry cheesecloth. Torretta puts on the paste (Fig. 3-13).

Step 4: Use a large brush when polishing. Don't be afraid to apply pressure. You won't hurt the furniture, and you'll get to all the waxed areas. Even a toothbrush will work (Fig. 3-14).

Step 5: Use more cheesecloth for a final polish, and brush (Fig. 3-15).

Extra hints from Torretta:

Clean brass tacks while they're on the chair, using steel wool. You can use steel wool to clean other brass accessories, but Torretta warns against rubbing too hard because you may wear away the plating or bend a piece out of shape.

An alcohol and shellac mixture (one part alcohol to two parts shellac) will bring up a bright shine.

Work on a chair in an upside-down position so you don't risk dripping on the seat.

Correct Warping

Richard Moller apprenticed at age 20 as a cabinetmaker because he had an avid interest in antiques. His uncle was art director of the Smithsonian for 25 years, his grandfather was a Danish clockmaker, and four other uncles were jewelers. Now he is the one who travels to museums and does one-site restoration. His most recent work was at the Phipps estate in Old Westbury, Long Island. Moller deplores the fact that there is no formal education for serious cabinetmakers in the United States.

Fig. 3-13. If you apply paste wax and the finish becomes sticky, it has an oil base. This means you must start over. Courtesy *The Family Handyman*.

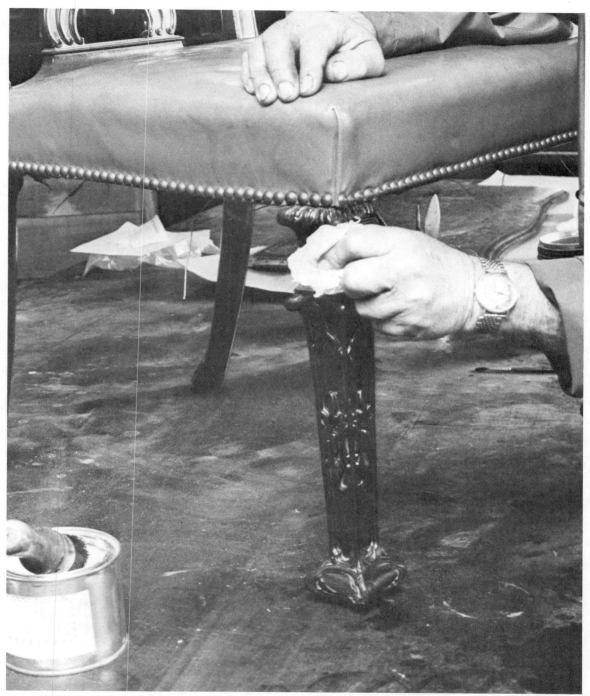

Fig. 3-14. Use a stiff brush to get into hard-to-reach areas, and don't be afraid to apply pressure. Courtesy *The Family Handyman*.

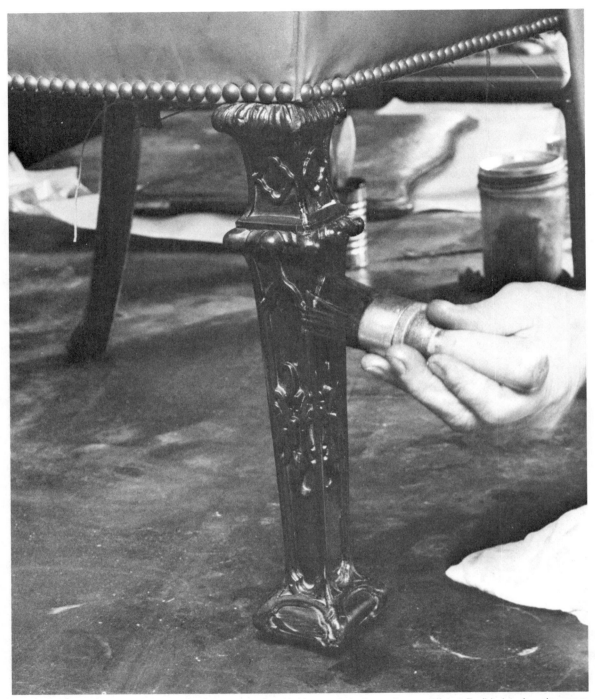

Fig. 3-15. After you've applied the wax, go over it with a dry cheesecloth, then a large, stiff, but flexible brush, using more cheesecloth for a final polish. Courtesy *The Family Handyman*.

Eight years ago he went looking for a job in restoration and remembers, "I went to about 50 shops, and there wasn't anyone under 55 working there. What makes Sotheby's Restoration unique is the youth of its workers because it is an incredibily difficult field to break into." Moller is now a master cabinetmaker and shop foreman.

The secret to correcting warping is to properly cut the grooves, according to Moller, who demonstrated the technique using a sample board. "The rule of thumb is to make your cuts 1½ inches apart if you have a moderate warp of up to 1½ inches out of line at the center. Make cuts closer together if you have a more severe warp. If you don't make enough cuts, you run the risk of snapping the wood when you clamp it.

"The other mistake many people make is not making the cuts deep enough. Don't be afraid to go two-thirds of the way through the board," says Moller. "You can run into problems if the wood is cracked, but you should be on the alert for this type of problem before you start cutting."

The Expert's Instructions

Step 1: Remove the piece of wood to be worked on . Turn it upside down. Do all the work on the bottom so the repair work isn't visible from above. This technique works for both solid wood and veneered wood.

Step 2: Cut the grooves with a portable circular saw. Clamp a fence for the saw to ride on, and then after each cut, move the fence to the next appropriate space (Figs. 3-16, 3-17).

Cautionary note: Using a circular saw (or any saw) can be dangerous. Use extreme caution, observe all safety regulations, and follow your manufacturer's instructions. Adds Moller, "Also keep in mind you are cutting through old wood. So, if possible, use a carbide blade."

Step 3: After cutting the grooves, thoroughly wet down and soak the wood before you clamp it.

Step 4: Clamp tightly enough so the warp is pulled out.

Step 5: Next, plane and fit shims into the grooves. Note: Shims for correcting warping are top-beveled (an inverse wedge that is thinner at top than bottom). This is so that it makes a tight glue joint when pushed flat.

Moller is a keen advocate of using a wood plane. "It's difficult to get used to if you've been using a steel plane," he says, "but it's worth it because, unlike a steel plane, it won't leave marks on wood furniture."

Step 6: Glue the shims into place and clamp straight. Let dry. Figures 3-18 and 3-19 show Moller at work.

Shimming Cracks

Tom "Tex" Tolson started working at the tender age of six as a plumber's helper to his grandfather. He claims to have learned "two dozen trades" before he began working in restoration at the advanced age of 17. "I'm so hyperactive, it's hard for a guy like me to have a job where I'm sitting down," says Tolson, who bicycled to New York City from his native Texas. In his spare time, he's an inventor. He holds a patent, with his sister, on a special storage door/cabinet design for space utilization and has recently invented a machine to remove graffiti.

It's a good idea to think about shimming a crack rather than filling it any time it's over 1/32 inch wide, advises Tom. "It may take a little more time, but you should be able to make and set a shim in half an hour."

The Expert's Instructions

Step 1: Clean the crack out. If the crack has already been shimmed, remove the old one

Fig. 3-16. To correct warping, be sure to properly cut grooves, fencing into place over the inverted piece of wood to be straightened. Courtesy *The Family Handyman*.

Fig. 3-17. Moller advises not to be afraid to cut a groove two-thirds of the way deep. Here he measures the depth and sets his saw. Courtesy *The Family Handyman*.

Fig. 3-18. For best results, reposition the fence and clamp for every cut. Courtesy *The Family Handyman*.

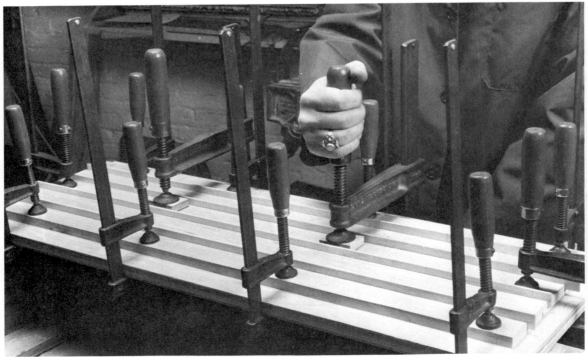

Fig. 3-19. The shimmed board is clamped down tightly so the warp is straightened. Courtesy *The Family Handyman*.

Fig. 3-20. To clean a crack, use anything that does the job. A hacksaw blade that has a handle made from a wooden dowel is good. Courtesy *The Family Handyman*.

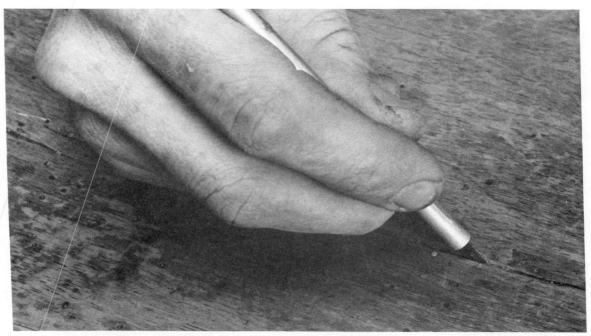

Fig. 3-21. Use a small chisel, X-Acto knife, dental pick, or even a jeweler's file to get into very fine or very tight areas of the crack. Courtesy *The Family Handyman*.

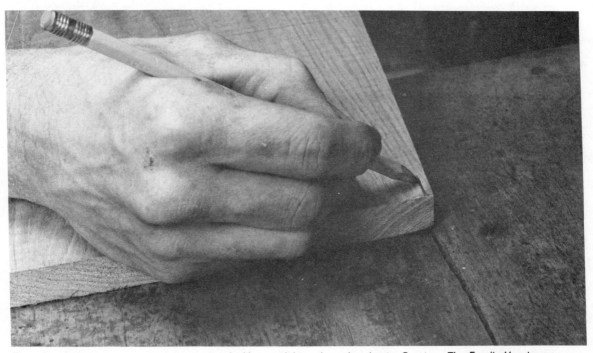

Fig. 3-22. When making a shim, choose wood with a straight grain and no knots. Courtesy *The Family Handyman*.

and make a new one to fit the entire area. If it's a jagged break, even out the crack so you have two straight sides for good contact when you get to the gluing (Fig. 3-20).

Tolson recommends using a hacksaw blade for fine shims (sharpen or feather the ends of the blade to help you get into the cracks). For even finer and tighter areas, he recommends a hacksaw blade where the set has been ground off. "You can use a thin rasp for wide cracks as long as you keep in mind the object is not to distort the straight lines," Tolson points out. The basic idea is to use as many blades as necessary to thoroughly clean the crack. Get into the crack (Fig. 3-21).

Step 2: Select a piece of wood to make the shim (Fig. 3-22). Ideally, you want to match up the type of wood. You also want to get a shim that has a straight grain (wild grain will snag the plane) and is knot-free (sawing into a knot can shatter the shim).

Step 3: Cut the shim roughly to size using a saw. Try to cut it both wider and thicker than needed. You want it deeper than needed as well. If you cut the shim precisely to size, you can lose part of the shim when you hammer it in (Fig. 3-23).

Tolson advises that if you need a thin shim, you should push it through a band saw. Use a table saw only for cutting a thicker piece of wood.

Step 4: Plane the shim to fit. Check the grain before planing, then clamp one end of the shim to the table (clamp on wide side of wedge). Figures 3-24 and 3-25 show shaping up the shim.

Plane in the direction away from clamp. Cutting the shim to fit is mostly a matter of trial and error—first planing and fitting and then planing again. You might mark the back side of the shim with a pencil for different high spots so you don't plane down too far. You may also want to use short strokes if you can't drag the plane the length of the shim.

Step 5: Sand and shim. This will give you an even tighter fit (Fig. 3-26).

Step 6: Glue the shim into position. With a plastic syringe, squirt the glue into the hole and spread it with a flat blade (such as a pallet knife or kitchen knife). Work fast so the shim won't stick. Then apply glue liberally to the shim (Fig. 3-27). Wipe off the excess (don't be afraid to use your fingers).

Step 7: Drive the shim into the crack with a wooden mallet. Tap, don't hammer (Fig. 3-28).

Step 8: Trim off any excess shim with a chisel or plane until it is flush with the surface (Fig. 3-29).

Step 9: Dry the area by wiping it with cheesecloth that has been moistened with warm water. This removes any excess glue. Now dry the area with a paper towel.

Step 10: Sand the shim down. Tolson puts masking tape on all sides of the shim to minimize the risk of damaging the finish while sanding. It also feathers the tape in.

Step 11: Fill hairlines and step shoulders. Use a paste of same-wood sawdust and glue. Apply the glue in a circular motion rather than going with the grain. This removes excess paste.

"You may want or need a second coat because the glue may shrink as it dries," notes Tolson.

MORE TIPS FROM THE EXPERTS

Here are some general restoration tips from the pros at Sotheby's.

1. "Don't throw away your sawdust. Save it in paper cups or film canisters and use it to fill in screw holes. Belt sanders produce a finer dust than band saws. If you don't have the same-wood sawdust, mix different ones together to make a color you are missing." (Tom "Tex" Tolson.)

Fig. 3-23. Cut the shim roughly to size with a saw. Courtesy *The Family Handyman*.

Fig. 3-24. Planing the shim to fit is mostly a matter of trial and error. Courtesy *The Family Handyman*.

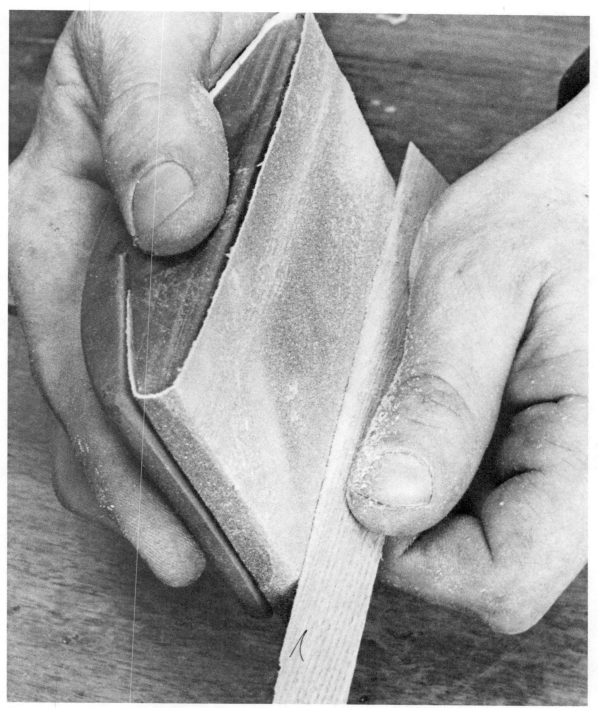

Fig. 3-25. After planing the shim, sand it for an even more perfect fit, remembering patience. Courtesy *The Family Handyman*.

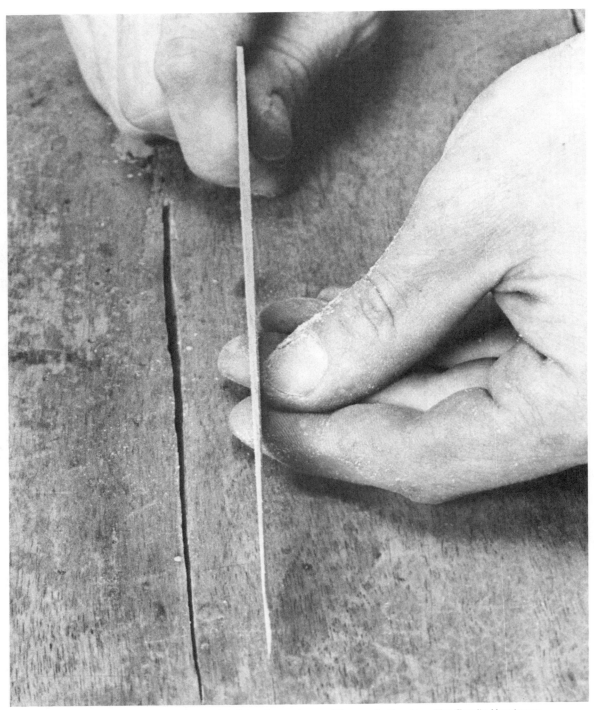

Fig. 3-26. A few of the finished shim notes it follows the contours of the crack. Courtesy *The Family Handyman*.

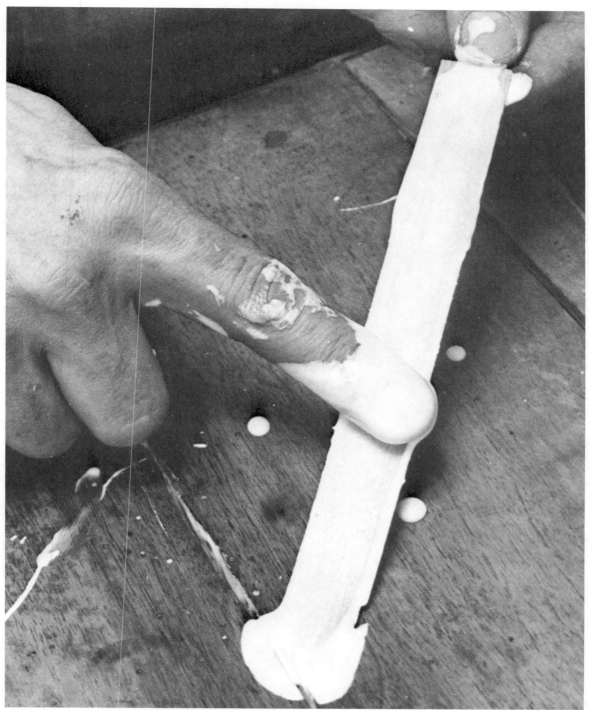

Fig. 3-27. Tolson demonstrates just how liberal you should be when applying glue. Courtesy *The Family Handyman*.

Fig. 3-28. The shim is now carefully worked into the groove with a wooden mallet. Courtesy *The Family Handyman*.

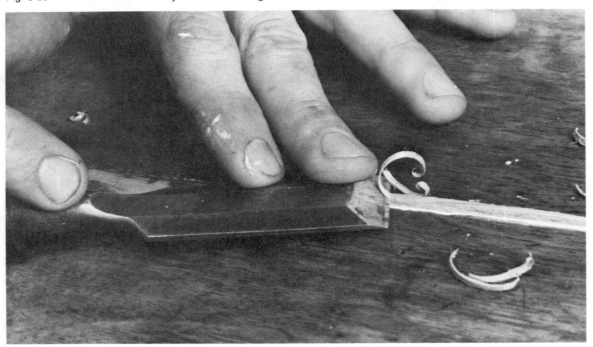

Fig. 3-29. Chisel or plane off the excess from the shim until it's flush with the surface. Courtesy *The Family Handyman*.

2. "Protect the wood you are working on by covering your workbench with a blanket or even a towel, especially when gluing." (John "Scottie" Scott.)

3. "Avoid unsightly temporary repairs, such as nailing on a loose leg. It's what I call 'human termite error'—people do more damage than time or bugs. Only gluing will really repair a loose leg. Makeshift repairs now will only have to be redone sooner or later." (Alan Hamilton.)

4. "When clamping, be careful not to outweigh the piece you are working on. Steel clamps are very heavy, and their weight alone can actually create warping. Therefore, when clamping, use the lightest weight clamp possible that will do the job." (Tom "Tex" Tolson.)

5. "When cleaning and polishing furniture, don't kill yourself trying to remove all of the dirt and grime. You want to leave some of it on. The idea isn't to make the piece look new, but to leave the character it has acquired over the years. We clean the center more than the edges to achieve this effect. Also, you don't want to polish it to death." (John "Old John" Torretta.)

6. "Staining and polishing will only hide tiny imperfections. Anything else will actually become more obvious. For truly professional results, aim as close to perfection as possible." (Alan Hamilton.)

7. "Use wax a shade darker than the wood. This will prevent the wax from turning white in the crevices. (John "Old John" Torretta.)

8. "Take some advise from a guy who learned the hard way. Think out ahead what you are doing, especially when using a saw. Never be in too much of a hurry or feel too sure of yourself to ignore basic shop safety. My mangled finger is the result of a run-in with a table saw, a vicious little machine that doesn't stop for anything—especially flesh and bones." (Alan Hamilton.)

9. "Cracks too small to shim can be filled with wax." (Tom "Tex" Tolson.)

10. "Use spackling compound instead of gesso. It's more flexible, and it sands more easily when you have small cracks in lacquer." (Allison Armstrong.)

Thus ends an excellent course in gluing and working with fine furniture, old or new.

Chapter 4

Plumbing Repairs:
When Glue Goes on the Line

As this page is being written, the nation is pretty well frozen, including Florida. Below-zero temperatures have been the norm for some two weeks. Hundreds of thousands of pipe lines are broken due to the freezing. This book is being written in Asheville, North Carolina, and some 30,000 houses have burst water pipes.

An adhesive or sealant that will repair plumbing problems is extremely important. Most of the pipes spewing water into homes and basements this minute could be temporarily fixed with a good plumbing repair sealant.

The finest we have found is Hermetite's Leakstopper. It stops leaks instantly, is waterproof, and is not affected by gas, oil, or alcohol. It seals leaks in pipes, tanks, radiators, and couplings, and has a multitude of other uses. It will adhere to copper, steel, lead, cast iron, ceramics, and many other surfaces.

For the best results, smooth surfaces should be roughened with sandpaper or a file before the Leakstopper is applied. The tremendous Hermetite product is very easy to use. You simply break off equal portions of the resin and hardener, then press and knead the putty in the palms of the hands. After it is well mixed, roll the putty between the hands. Continue the kneading and rolling until the streaks disappear. When the streaks are gone, the material is ready to apply to a clean surface.

"Leakstopper remains workable for up to two full hours," Frank R. Noel, the president of Hermetite Products Inc., says. "During this time after mixing, it may be molded or formed to any desired shape. Since it sets rock-hard, it may be drilled, filed, sanded, or painted. You may obtain a very smoothly finished surface by gently rubbing with a wet finger or damp cloth immediately after the application of the Leakstopper."

Repaired areas can be painted immediately after the Leakstopper is used. You should wait at least three hours before applying any water-based paints. Leakstopper reaches a full cure in 24 hours. After that time it can be drilled, sanded, or machined to desired look. Although the Leakstopper is resistant to chemicals, thin layers may be softened for removal using a proprietary paint stripper.

Another outstanding asset of the Leakstopper is that it is electrically nonconductive. It is also an ideal compound with insulating properties. It works under water, making it one of the most versatile plumbing products for a variety of home uses. Usually, there is no need to cut off the water supply when making repairs with the Leakstopper. It is always best to have the area dry, though. When possible, the water should be cut off while making the repair.

INSPECTING YOUR PLUMBING

Check your plumbing for any leaks, especially around joints. Trap area under sinks should be examined to ensure that no water is seeping out. The kitchen sink needs a periodic examination also. Don't forget to check any outside lines leading to faucets, and washing machine connections should be examined for any breaks.

The quality of Hermetite Leakstopper is so superb it has no competitive counterpart! Better hardware, home center, and other stores stock it. There are some products that claim to be equally effective, but they are not, according to research reports. If you are unable to purchase the little packages in your area, you should contact Hermetite Products Inc. at 6500 Glenway Avenue in Cincinatti, Ohio 45211. No home or office should be without it.

Buckets, cans, and other containers that leak can be repaired with Leakstopper. Cracks in almost any metal can be sealed with the compound. Toys, including bicycles, can be repaired in many instances with the product. Older gutters that need repairs can be fixed with Leakstopper. It is excellent for repairing the drainspouts on gutters. It is also the solution to leaks in most metal gutters (Fig. 4-1).

CAR LEAKS

Leakstopper is a necessity outside the house as well. The plumbing lines in recreational vehicles are as susceptible to problems as those in any other place. The adhesive should be in the repair supplies on any RV. Boats with traditional plumbing also need it on hand. It definitely should be in the trunk or glove compartment of every car. The Hermetite adhesive has a host of important automotive uses.

Cars are leaking by the millions right now. Quite often, it is difficult to find where a leak is coming from. The fluid coming out is a strong lead to where the leak is located. Examine the liquid coming out of your car. Dark, smelly, slick liquid is usually oil coming from the engine, the transmission, or the oil drip pan. Rusty or greenish, slightly scented fluid that is rather slick is probably antifreeze coming from the cooling system. It is usually green when it first leaks out, but it tends to turn rather rusty in a few minutes.

Any clear, odorless, and non-slick fluid is probably water coming from the heater, air conditioner, windshield sprayer, or radiator system. In cold weather, some water will appear from the tailpipe when the motor is first started. This is just water coming from the gasoline in an internal combustion engine. It is just a byproduct.

A yellowish, rather pungent-smelling fluid directly under the battery is battery acid, and you should avoid getting it on your skin. Honey-colored or red liquid which is not as

Fig. 4-1. Gas cans and other metal containers can be repaired with Hermetite's Leakstopper. Courtesy Hermetite Products Corp.

slippery as oil is probably automatic transmission fluid coming from the reservoir. Another honey-colored liquid that says you have problems does not smell; this may be coming from the steering column section (hydraulic fluid) from the master cylinder, wheel cylinders, or brake lines (brake fluid). The stuff is corrosive, so contact with the skin should be avoided. A rather clear liquid that may have a tinge of yellow and a strong smell is probably gasoline, especially if it evaporates quickly. It may be coming from the tank, the gas filter, the fuel lines, or carburetor. Leakstopper can solve many of these leaks. This could be of the greatest help when sitting beside the road far from help. Even if you don't have the mechanical knowledge to fix the leak, if you have some Leakstopper with you, you might find someone who can make the repair, saving you much time and maybe money. At the minimum, the amazing adhesive might save you a towing bill.

Leakstopper should be kept in second homes or vacation cottages also. Going to stay for a time in such a place without having some Leakstopper around could really ruin a vacation or outing.

RIGID PLASTIC PIPE

PVC (polyvinyl chloride), ABS (acrylomitrile-butadiene-styrene), and CPVE (chlorinated polyvinyl chloride) are the three types of rigid plastic pipe used in home plumbing. Building codes usually restrict these to being exposed, meaning they are not to be used within or behind walls. They are primarily for cold water use and drainage systems. CVPC is an exception which may be used for both hot and cold lines in some cases.

Plastic pipes can be joined to plastic fittings with a solvent cement. Before using the solvent welding, put the pipe and the fitting together, making sure the pipe slides all the way into the fitting before it touches its shoulder. This testing for a perfect fit is important because the solvent sets too quickly for any major repositioning.

Always check the solvent, and discard it if it has thickened or discolored. CPVC solvent weld comes in two containers, one being a cleaning agent. ABS and PVC solvents come in one can. All types of solvents should be applied only with brushes with non-synthetic bristles. With the CPVC, use a separate brush for the cleaner. Solvent-welded joints should be allowed to dry for at least three minutes before

moving them. No water should be introduced into the pipes for at least 16 hours after their joining.

The strong plastic pipe can be coupled tubing with flare fittings. The pipe end should be warmed with a torch before flaring, and there are special adapters for making plastic-to-steel connections. An adaptor threads on one end to accept steel or brass pipe, and the other end has a plastic bushing that cements to plastic pipe. Don't mix different types of rigid plastic pipe, such as an ABSA pipe to a PVC fitting, since it will probably leak.

To solve-weld a joint, the plastic pipe should be sawed in a miter box with a 24-tooth blade hacksaw. This gives the smoothest cut, much cleaner than an ordinary handsaw. Run a knife inside the pipe to remove the burr. A file or sandpaper will also work well. Use fine sandpaper to get the best surface. With CPVC pipe, the cleaner is used first to remove dirt, oil, or wax, applying it to the outside of the pipe where it joins and the inside of the ring. The solvent from the second can is then brushed liberally on the outside of the pipe and lightly inside the fitting.

For the other two types of plastic pipe, this is only a one-step process. After coating them, immediately put the pipe and fitting together with just a quarter or so twist until the pipe seats. Quickly move the fitting to the desired position. When the match is good, there will be a bead of fused solvent around the joint. If the joint leaks when you try it a day later, it will have to be cut out and replaced with a new coupling or fitting.

PIPE JOINT COMPOUNDS

Coating pipe threads with a joint compound is standard to plumbing procedures. They help seal joints and make it much easier to take joints apart when repairs are needed. Teflon tape is also being widely used in place of the compound. Some pipe sealants also contain Teflon. The best one will cure pretty well in one day, but three days are best for the full cure. These sealants have to harden without any shrinking to give the strongest bond. They are very hard and make for a strong plumbing system, resistant to vibrations that may tear loose weaker connections.

In contrast to these anaerobic sealants, there is a class of pliable sealants which allows for expansion and contraction of pipes due to pressure and temperature changes. They move and conform to the pipes, sealing any shifting of the system and filling gaps due to strains. Some of these are appropriate for pipe systems that must accommodate temperatures 50 degrees below zero or up to 400 degrees F. These are primarily used in industrial applications because they are resistant to most solvents.

Chapter 5

Glues and Adhesives in Building Construction

Adhesives are rapidly replacing nails and other fasteners in the construction world. Walls, floors, and ceilings are often glued into place, and they are stronger than any other means of keeping them in place. Gluing is also the fastest of all fastening systems in most projects.

PANELING

Millions of people put up their own paneling when building a new home or remodeling. Before remodeling the walls of a room, it is advisable to shop for paneling with great care. The remarkable selection demands the utmost attention. More than 50 wood grains are available, and many of them take a very trained eye to determine that they are not real wood all the way through. There are also many panels on the market with effects other than wood, such as veined marble or hand-hewn stone.

Once you determine how much paneling will be needed, have it delivered to the room where it is going to be used several days before you start the project. This will expose the panels to the room's normal humidity and allow them to adjust to it. This is important regardless of how the panels are going to be fastened. When the panels are stacked, thin sticks should be placed between them so the air will circulate well around them.

For random-width paneling, the furring strips should be applied vertically at 16-inch intervals, and horizontally at 16-inch intervals over the existing two-by-fours. Precise measuring is important to allow for light switches and other openings. Set the panel in place before applying the adhesive. If it fits well, then apply the adhesive at the 16-inch intervals, horizontally and vertically. Apply beads of adhesive to all the furring strips that will come in contact with the paneling. If needed, put some cleats between the strips in areas where the paneling should be particularly strong or to support especially heavy paneling.

Next, position the paneling against the furring and drive a few small nails near the top of the paneling to create a hinge-like action. It is also good to press the panels against the strips so they will pick up some of the adhesive from the strips. Pull the bottom of the panel away from the wall about a foot—10 inches is a good measure—and prop out with a wood block. This keeps the bottom part separated from the strips, allowing the adhesive time to develop the needed tackiness.

Within eight to 10 minutes for most adhesives, the time is right. Remove the block and push the panel in position against the strips, being very careful of the alignment. Cover a wood block with a cloth and tap the sheet to spread the adhesive evenly. When the adhesive has set securely, you may remove the nails at the top or sink them. Many people find it is good to nail the panel at the bottom also, giving the panel the greatest adhesion to the wall.

Macco's Liquid Nails® adhesive is one of the best construction adhesives. It is a high quality waterproof general-purpose adhesive formulated for bonding prefinished panels of all types—plywood, hardboard, drywall, and similar materials—to wood, steel, plaster, concrete, and most other common building materials. Liquid Nails is made of virgin rubber and will retain its bond and flexing properties for absorbing wall movement due to shock or normal settling years after other products have become brittle. It requires no mixing and has good open time. It is also good for installing brackets, storage walls, door and window trim, towel bars, veneer brick, and replacing loose ceramic tiles. It has a very low odor since the solvent emits no offensive smell.

When bonding panels directly to wall studs, use the 11-ounce Liquid Nails cartridge (LN-601) and a commercial caulking gun. Gun a ¼-inch bead of adhesive on each stud to be contacted by a panel. Push the panel firmly into position and adjust.

There is also the contact method of applying the panel. After the panel is positioned, it is removed and the adhesive is allowed to air-dry for five to seven minutes. Then the panel is repositioned carefully, and pressure is applied.

To bond foam insulation between furring strips, apply the ¼-inch bead to the back side of the foam, which is cut to fit, the beads running approximately eight to 10 inches apart. Press firmly to the wall. When using polystyrene foam, the surface temperature should not exceed 90 degrees F. Avoid prolonged contact with skin and breathing vapors. Open all windows and doors, using as much cross-ventilation as possible.

Macco makes a Professional Liquid Nails (LN-602) which is an interior-exterior grade construction adhesive. It is a waterproof synthetic rubber and resin based product designed for interior or exterior use in installing wood decks using treated lumber), tongue-and-groove subflooring, adhering wall paneling, bonding and sheathing, applying lap siding and sheet siding, laying plywood subflooring, and setting prefinished doors, windows, and trim. It bonds to and remains elastic on most any construction surface including wood, concrete, steel, and galvanized metals and aluminum. LN-602 is not for use with polystyrene foam.

The LN-602 provides better stability and noise control than nailed floors, with less deflection, sagging, bouncing, or squeaking. Contractors using it have experienced drastic reductions in callbacks to repair squeaks and other troubles. The adhesive has been used effectively on wet or frozen lumber as long as the surfaces have been clean of snow, ice, and mud. It conforms fully to HUD-FHA *Use of Materials Bulletin UM-60*. It also passes American Plywood Association tests.

The product is good for factory-built modules due to its high bond strength. The modules may be transported over long distances and rough terrain to final sites without loss of tightness, even after the stresses of crane handling.

The American Plywood Association's bulletin *APA Glued Floor System* is recommended for guidance in construction. LN-602 is extremely flammable and care must be taken in keeping it away from heat, sparks, or open flames. Contact Macco Adhesives for answers to application questions at (800) 321-3647; in Ohio, call (216) 943-6161 collect.

Macco's LN-603 Liquid Nails is a trowel grade for the installation of decorative tileboard and other wall coverings. Macco's C-5 ceiling tiles, Macco's CT-20 ceiling tile adhesive is very good. Macco's TA-605 is for decorative tileboard applications. Macco also has DW-24 for drywall adhesions.

LAMINATES

Plastic laminates for home use are invariably attached with adhesives. The thin decorative sheets may be used in almost all flat and dry surfaces. Door, cabinets, and counters are among the most popular places where laminates are used. Generally, the surface to be covered should be roughened a bit so the adhesive will grab sufficiently. Working with laminates requires the utmost attention to measuring and fitting.

Some adhesives do not allow much time to position the laminate once it makes contact with the tackiness. Measuring carefully is instrumental to doing a neat job that will fit well.

The laminate should be cut a little too big to allow trimming when possible. Apply the contact adhesive to the laminate evenly. The other surface must also be evenly coated. Most adhesives recommend drying for about 15 minutes before the surfaces are mated. The strongest adhesives do not allow for any removal once contact has been made.

Another method of ensuring perfect alignment is to push some thumbtacks into the table top (or whatever is being covered) to be used for guides in lining up the laminate. Put some brown paper down on the dried, cement-coated surface. Align the dried, cement-coated laminate on the surface. Then withdraw the paper, pressing down on the laminate sheet as it contacts the surface to which it is being applied. Edging strips or moldings can be attached in the same way. Being as neat as possible with the adhesive will give you a much better job and not complicate your life in the cleaning afterwards. Some laminates and plastic coverings are applied to walls.

Macco's CA-90 nonflammable contact adhesive is a brushable grade. An even coat is applied to each surface, but plywood and other porous surfaces may require two coats. Allow the first coat to dry before putting on a second coat. The thickness may be gauged by obtaining a uniform, semi-glossy look to the adhesive film. When it is becoming dry to touch, the substrates may be bonded merely by placing them together. Parts must be positioned carefully, because the work cannot be shifted after contact is made. Use a mallet or rubber roller to apply pressure over the entire area, beginning in the center and working out. This avoids entrapment of air bubbles and gives best results. Clean tools and any spillage with the CS-96 Nonflammable contact Adhesive Solvent.

Ample working time is granted with this

adhesive since the substrates may be apart for 90 minutes before contact is made after they are coated. Its versatility allows it to adhere to wood, metal, leather, canvas, and other materials, provided both surfaces are always coated.

The compound based on neoprene rubber was developed mainly for adhering high-pressure plastic laminates to plywood, particle board, flakeboard, or metal. Bonding is instanteous upon contact, and high shear strengths are developed after curing.

Various brick, tile, stone, and other veneers are popular. Be very sure when buying any materials to be applied with adhesives that you are getting the highest grade adhesive. On occasion, some expensive materials have poor adhesives with them, especially when the adhesives are packaged with materials. When you are allowed to buy the adhesives separately, you may be more assured of getting the best adhesive for the job.

Some people find that using a damp cloth on the back of brick veneer moistens it enough to delay the setting time, giving you more time to position each piece precisely. Ceramic wall tiles require equal concentration on working with special adhesive. Ceramic wall tiles require equal concentration on working with the special adhesive. Ceramic jobs are often branded "mud" work. The mastics used on floors and walls are usually especially heavy. They are heavy-duty adhesives for heavy-duty applications.

Most mastics are rubber-based. They have needed flexibility to allow for movement while tiles are being set. It is best to follow the directions for each type of adhesive you use to the letter. Be sure not to cover too large an area with the adhesive. Manufacturers warn of this infraction often. Pressing the tiles into the adhesive properly is very important. When you

have finished an area, the adhesive residue should be removed before starting on the next section.

FLOOR COVERINGS

Tiles are popular on floors, and there are many types of other coverings. Resilient flooring—such as linoleum, vinyl, and tiles made of vinyl, vinyl-asbestos, and cork—is the most common. More and more people are becoming concerned about bringing toxic materials into their homes. Asbestos is related to numerous ills, and formaldeyhde is rampant in plywood and particle board. Cork is the only natural resilient flooring. The "wood tiles" are usually 5/16-inch fine wood (usually oak) attached to a foam base. A special adhesive holds them down. For old or uneven floors, an underlayment makes the ideal surface to apply any new flooring, and it makes the adhesive easier to apply.

In remodeling a room, damaged wall panels or floor tiles may have to be replaced, or some materials may need to be removed to put in new decorations. To pry off an old panel, start with a putty knife and then work up to a wide, flat chisel or small pry bar. Care must be used in removing any moldings at the top or bottom of the panel. Getting the old glue to turn loose may be a war. When all of the panel is removed, the old glue will have to be scraped off the studs or strips.

Once the area is cleaned, put the new panel in place to check it for fit. If the fit is fine, then the adhesive should be applied to the panel backing and the wall carefully. Be sure not to use too much of a bead near the ends of the panel, so there will not be an excess to cause cleaning problems once the panel is secured.

The easiest way to replace a damaged floor or wall tile is to remove the old tile without damaging any others. A torch or hot iron will usually loosen the old adhesive, enabling you to lift the old tile out. The remainder of the adhesive must be removed. Make sure the new tile is going to fit well. If the tile presents any problem in fitting correctly, it may be warmed with an iron or torch until it is somewhat flexible. Apply the adhesive in the right amount according to the directions for the type of tile being used. When the replacement tile is in place, weight it down for the total drying period.

One of the finest multi-purpose flooring adhesives is Macco PV-20. It is suited to both the professional and the home handyman for installing most popular types of soft and resilient floor coverings. It performs well with tiles of solid vinyl, vinyl laminate or rubber, jute or rubber-backed carpet, and sheet goods. Macco also makes a vinyl asbestos and asphalt tile adhesive, BL-70.

M-112 from Macco is a ceramic floor tile and gauged slate adhesive. It sets mosaics, quarry tile, pavers, and slate. It is very versatile and handy for repairs. The M-221 ceramic floor tile adhesive is similar in its uses.

One of the most unusual adhesives in the Macco line is the RC-711, a releasable carpet adhesive. It allows you to remove the carpet for cleaning and or replacement. Another strong asset is that it does not require a primer. This saves much time and money.

The SB-60 carpet adhesive is superior for indoor-outdoor carpeting and foam or sponge rubber-backed carpeting. It has a fast tack and lasting bond over dry, smooth, well-cured concrete, wood, and hardboard. It is freeze-thaw stable and may be used in any room or protected outdoor area appropriate to the carpet being used, where the surface, carpet, and adhesive container can be kept above 70 degrees F for 24 hours prior to working.

WALLPAPER WITHOUT A WRINKLE

Hanging wallpaper can be one of the most frustrating and challenging home projects for do-it-yourselfers. It can also be almost as easy as the directions in the little booklets down at the wallcovering stores. Removing the old paper may be the biggest battle. There are various ways of stripping off the old paper. They include scraping, steaming, and the use of commercial solvents that will persuade the old paste to let go.

There are many methods of applying the paste. Vinyl or Mylar wallcoverings that are not prepasted may be coated with a paint roller, or the paste may be put directly on the wall. Wallpapering demands that you have the right tools. A two-foot level, a paste brush, a smoothing brush, large sponge, paint roller, tape measure, utility knife, seam roller, scissors, joint knife, and straightedge should be on hand. A good premixed vinyl paste, a dropcloth, and plastic bucket complete the necessities.

The paste is more important than the paper. Poor paste will not adequately hold the most expensive—or cheapest—paper. There is a wide range of wallpaper pastes. Make sure you use the best. Devoting the proper amount of time to each phase of papering will ensure the best job. After some strips are in place, wait a couple of minutes for the paste to begin setting before putting the roller on the seams. If you use the seam roller too quickly, you will squeeze out too much of the paste, reducing the strength of the strips to the wall. Just getting out the excess paste will give a stronger bond and also reduce your clean-up time. The highest quality paste shows its properties in corners and tight spots, where cutting, trimming, and peeling portions of the paper away from the wall must be done.

HEAVY-DUTY
WALLCOVERING ADHESIVE

The finest grade of wall adhesive comes from Associated Paint & Plastics Corporation at 7575 N. W. 74th Avenue in Miami, Florida 33166. The product is called Heavy Duty Wallcovering Adhesive, 80-500. It is a water-based, heavy-duty, ready-mixed paste for hanging vinyl wallcoverings of all weights. It forms a tight bond on new or previously painted plaster, drywall, wallboard, masonry, wood, and metal. It resists water, staining, and bleeding to adjoining areas.

In compliance with federal regulations, it has less than .06 percent of lead in its non-volatile ingredients. It also conforms to air pollution regulations in its emission of hydrocarbons. It is not recommended that it be applied when the air or surface temperature is below 50 degrees F.

The surface preparation for use of the Associated Paint & Plastics adhesive is standard. Remove any grease, dirt, loose paint, and any crumbly or powdery material. Sand any slick or glossy old paint surfaces. Any mold or mildew must be removed, and the surface should be sterilized prior to coating. If 25 percent or more of the previous coating has failed, complete removal is highly recommended.

All metal surfaces should be primed prior to coating, and in cured plaster the moisture content should be less than 15 percent. Prime with 45-650 Vinyl Primer Sealer on new drywall, and use 80-510 Prep All on previously painted drywall. Prime wallboard with 80-510 Pre All; this works well on masonry also. On new wood, prime with 45-500 Enamel Undercoater; on old wood, sand any glossy surface and prime with 45-500 Enamel Undercoater. It may be applied with brush, roller, or machine coater.

The recommended spreading rate of the Heavy Duty Wallcovering Adhesive is 100 to 150 square feet per gallon, depending on surface texture and porosity. Open time is 20 to 30 minutes depending on temperature, humidity, and the thickness of application. Wash excess paste from seams and areas not to be covered with clean water before the paste dries. The adhesive may be cleaned from the tools promptly with warm soapy water.

According to many wallpapering authorities, this wallcovering adhesive is the finest. It is not yet available in all states. If you cannot obtain it in your area, write to Associated Paint & Plastics Corporation at the address noted at the beginning of this section.

Remembering the basics about wallpapering will make the job much easier. It is important to use a clean brush when applying the paste each time, so be sure your brush is always clean. The paste must be spread very evenly because unpasted spots will blister. Work from the top of the strip at the center, and leaving an inch or so unpasted at the top edge makes it easier to control, pasting this section last.

Shorter strips are easier to handle. Ceilings should have the paper run across the room rather than lengthwise. And, it is always best to do the ceilings before the walls. Corners are rarely true, and the quality of the paste is really tested with attention to trimming and fitting. With a little overlap in some cases, the room not being perfectly square in the corners is overcome, and it is not noticeable.

THE UNDERGROUND ARCH

Like other energy efficient structures, the underground house has strived to become one of the best. One of the most attractive and efficient of the underground types is the Arch-Tech panel made by Integrated Building Systems of Grand Haven, Michigan. Created by the two men who established the firm—John Loveless and Jeremy Berg—the curved four-foot by eight-foot panel is made of Ox-board. It is a type of particle board glued together with an adhesive made from ox blood.

These special panels are the essence of the Arch-Tech underground houses. They are bolted and glued to one another. The rather thin-skinned wooden arch is extremely strong. Earth on top of the house must be applied with precision. Equal amounts are put on each side until the proper balance is achieved. The weight of the earth actually bows up the arch a bit. (Loveless noted that his own home's arch rose less than an inch.)

Isoset is the adhesive that holds the unique arch together. This emulsion polymer isocyanurate is made by Ashland Chemical Company, Box 2219, Columbus, Ohio 43218. Resorcinols are used by many builders, but this special structural adhesive cures at a much wider range of temperatures, and Loveless and Berg are to be commended for selecting it because it is less toxic than most glues used in construction.

The top of the arch on the Loveless house has 18 inches of soil on it, while some 14 feet of earth cover the ends of the arch. Compared to conventional stick-built construction, the Arch-Tech panel house is no more expensive to build, and it may be built faster. This time factor also saves much money, and a shell can probably be built in the Western Michigan area (where the firm is located) for probably $13 to $15 per square foot. Although the company has concentrated primarily on its immediate region, it has homes from New Hampshire to North Carolina, and homes are being planned in Georgia and Montana.

The light weight of the panels (compared to most other building materials) is an asset in construction and in shipping. The panels may be sent anywhere in the nation for $500, the company says. The ox blood glued surface of the arch panel never comes in contact with the ground. Pressure-treated sheets of plywood are glued and nailed to the arch.

CLAMPS AND CLAMPING

Glue and clamps are usually synonymous. Setting weights on things that have been glued is also a form of clamping. The object is caught between the floor and the weight on it. Woodworkers have the most knowledge and devices for clamping. All joints require glue, even the most sophisticated ones such as the dovetail used in better furniture and fine construction. Knowing the basics about clamping can give you a better glue job every time.

The Versatile C-Clamp

C-clamps—also called carriagemaker's clamp—are appropriately named, since they have a C-shaped body with a bolt through the base. They range from tiny to eight inches or more in size. Since bare metal against wood may leave undesirable marks, putting in some thin scrap wood pieces to serve as jaw pads is advisable. These pads help distribute the pressure evenly.

Spring Clamps

These resemble battery jumper cable clamps. They are good for holding smaller glued parts together until they are dry. They come in a variety of strengths and sizes, but they are for lightweight tasks. They are especially suitable for fast-setting glues where several clamps may be needed to do a good job, though. The bigger clamps can handle some stronger jobs. These clamps are often underrated.

Hand Screw Clamps

These are favorites for woodworkers. Many have maple jaws, which require no pads and will not mar the wood they clamp. Hand screws adjust to many angles, and they are very strong. To be effective, the jaws must be parallel when a piece is clamped.

Bar or Pipe Clamps

Cabinet and furniture makers use these most often. They are generally sold without the pipe, which you must buy separately. They come in a wide selection of sizes. The vise jaws are already mounted on flat steel stock. To use them, set the fixed jaw against one side of the work and slide the movable jaw against the other side, then tighten the clamp with the hand screw.

Band and Web Clamps

To clamp irregular shapes and to draw together several joints at one time, canvas or nylon bands are best for the task. Gluing chair rungs is an example. Most bands are from 12 to 15 feet long and tighten with a crank or ratchet, depending on the type and make. This clamp also works well for gluing a drawer back together. Care must be taken not to break the things being glued, since the bands tighten with tremendous power.

Vises

Bench vises are ideal for clamping many items during the gluing process. Smaller vises will accommodate most small tasks; the bigger vises will do many jobs. If a clamp-on vise is being used, it is sometimes easier to carry the vise to the piece that needs to be clamped. Pads are needed to protect the object being glued from the rough jaws of the vise. The vises used by woodworkers have jaws lined with wood for protection.

Sometimes a combination of clamp types will be needed on some bigger or unusual jobs. It is best not to apply more pressure than absolutely necessary to hold parts securely. Too much pressure will just force out too much of the glue, resulting in a bad job.

Repairing and Replacing Shingled Roofing

Leaks can cause major damage to the insulation, framing, or ceiling below. Locating a leak in a roof can be very challenging. Quite often the leak in the roof is not directly over the wet or damaged area. If the house has an attic, finding the leak will be much easier. Careful inspection may find the bad spot quickly.

Even badly damaged and missing shingles may be replaced without having a major re-roofing job. Shingles are put on starting on the lower edge of the roof surface. You may have to lift the good shingles from above the damaged ones in order to slip the new shingles underneath to maintain the pattern.

Shingles must be warm enough to work them without cracking or breaking. Minor splits or tears may be repaired with a good roof cement, a trowel, roofing nails, and a hammer. Once on the roof safely, you must remove the nails from the damaged shingle and the one dirctly above it. The new shingle should be slipped under the raised shingles. Cutting off the corners may help get it into position. Don't peel off the paper strip on the back of the new shingle. Seal with an adhesive cement. Use roof cement to tack down the tabs on the row of shingles above the repair line. Curled or wind-blown shingles may be tacked down again with roof cement. Apply a small dab of cement to the nail heads also.

Slate roof shingles require more care in repairing and replacement. Removing the old shingle must be done with care. It is best to ask for advice or a sheet detailing the instructions on how to work with slate shingles at the place where you purchase the replacement slate. Roof cement must be applied to the roof area and also to the back of the slate. The nail holes must then be sealed when the slate is attached.

Asphalt Shingles

The key to repairing shingles is to get to the job before extensive maintenance is required. When you let this job slide, you are letting yourself in for damage to the insulation, framing, and maybe even to the ceiling below. If you exert a little care, damaged and missing shingles can be replaced without a total reroofing job.

Shingles are initially applied starting at the lower edge of the roof surface and running to the top in an overlapping pattern. Lift the good shingles above the damaged ones to slip the new ones under. This way you can maintain the original pattern.

Detecting the source of a leak can be frustrating. First carefully inspect the roof sheathing and rafters. If you cannot visually detect the leak, use a garden hose and run water over the suspected area. Be patient and allow the water to seep into a hole even the size of a pinhole. Any interruption or penetration of the roof is a suspect for leaks. Be sure that you check valleys and pipe or chimney flashing very carefully. Always remember that such leaks are difficult to locate at very best.

Warning: Never work on any roof when it is wet or windy. You are just asking for a major injury when you do so. Also, wear soft-soled shoes when examining the roof. This both keeps you from sliding and keeps the roof from being further damaged. While walking on the roof, do so in a crouched position to keep your balance and keep tools close at hand to prevent useless stretching and reaching.

Shingles loosened, ripped, or lifted by a storm are fairly easy to repair. Choose a warm day to do the job so that the shingle will be pliable and not too stiff. You will need roof cement, a trowel, roofing nails, and a hammer.

If the shingles are partially lifted or curled, you can repair them with just a dab of roof cement spread under the lifted areas with a trowel. To prevent further cracking, be sure that you choose a warm day for repairs.

If the shingles are badly torn and abused, generously apply roofing cement underneath. Press the cemented shingle down and nail the edges with roofing nails. Cover the nailheads with cement to prevent further leaking.

Sometimes repairs must be done immediately and you don't have an extra shingle. Use a sheet of aluminum or copper to make such a repair. Coat the bottom of the sheet with roofing cement and push it as far as you can under the damaged shingle.

Replacing Asphalt Shingles

The first step in replacing an asphalt shingle is to raise the shingle above the damaged one. Be very careful not to crack the good ones. Remember, they crack under pressure very easily, especially in cold weather. Remove the old nails with a pry bar or chisel. Slip the new shingle in place under the raised ones. If the shingle sticks and refuses to slip under the raised ones, rounding the corners may help. Be very careful not to tear the roofing paper. While holding up the shingles above, nail down the new shingle. Always use special roofing nails with broad heads, and always apply a dab of roof cement to the nailheads to prevent further leaks.

Hip Shingle Repairs

Hip joint shingles are no more difficult to repair than any other kind. Hip joint shingles showing cracks may be repaired successfully by using black asphalt roofing paint. If the cracks are wide or if the shingles are badly damaged, it is wise to use roofing cement.

Replacing Hip or Ridge Shingles

Replacing hip or ridge shingles is more difficult, but may be successfully completed with a little care and patience. The replacement hip or ridge shingles should always have a minimum of a three-inch overlap. Always nail down the bottom shingle at all four corners under the overlap. Be sure that you apply roof cement to the nailheads. The next step is to apply a coating of roof cement or asphalt paint to the bottom of the new ridge shingle before laying it down firmly. Once again nail and apply cement to the nailheads to prevent further leaks.

Repairing Slate Shingles

Slate shingles that just need repair are an easy job to handle. A cracked or broken shingle that is in proper position should be protected with a very generous coat of roof cement. Use a putty knife to apply the cement. Be generous and liberal in application.

Replacing Slate Shingles

You can make a nail puller out of a piece of strap iron. File two notches in it and bend the end over for a hammering surface. Now hook the tool over the nail, and drive it out with a hammer. If you drive the nail from inside the house to start with, it will help. Slates may be cut easily to fit by scoring each side with a chisel. If you don't have a chisel handy, use a screwdriver. Make sure that the scoring lines on both sides match.

Once the scoring step is finished, tap the slate along the scored lines to naturally deepen the marks. Over a straightedge, break the slate apart. Any ragged edges left may be smoothed by lightly striking them with a ham-

mer. Since slate is very brittle, it would be a good idea to practice a little first before you undertake a major replacement job.

Next, position the slate and mark the locations of the needed nails. Two nails are always needed to hold a slate shingle. Use a nail set to punch the nail holes through the slate. If you really want to be on the safe side, and are inexperienced to boot, use an electric drill. Apply roof cement to the missing slate areas and to the back of the new slate pieces. Push the new slate firmly under the upper slates. Drive in two roofing nails thorough the previously punched or drilled nail holes. Cover the nailheads with roofing cement.

SEALING FOUNDATION CRACKS

A patch material with a powerful bonding agent is needed to fix cracks in concrete or stucco siding. The damaged area must be cleaned with a scraper or wire brush to remove the old concrete that is loose and scaling. Wetting the area to be fixed with a liquid bonding agent is the first step. The patching material must be mixed precisely according to directions. The best adhesive patching will stick to even a feathered edge. A good masonry sealer should be applied to complete the repair. When properly done, the repair will not be very visible, if at all. Such patching repairs may be made above or below grade.

DECAYED STONEWORK

Small damaged areas in natural stones may be repaired with epoxy-based fillers that are made for concrete work. Natural stones usually do not flake as brick does. Air pollution is so high in some areas that it does damage stones, though. These special adhesives are even suitable for repairs to stone steps. Some of these special glues tend to be rather expensive. Stone and brick dealers can tell you how to make certain repairs, and they stock the adhesives need.

BLACKTOP DRIVEWAYS

Good blacktop fillers and sealers have enough adhesion to set strongly. Thoroughly cleaning the hole or depressed area on blacktop is very important. Deeper holes should be filled with loose gravel. Cold-mix asphalt materials work well. The patching compound should be warm enough to be pliable. You may want to warm the material elsewhere, and then bring it to the site. Fill the hole to within an inch of the top with the patching material. Tamping it with a block will get rid of any air pockets. Then finish filling the hole, mounding it over a bit until it is pretty close to being flush. Driving your car over the patching will get it flat. Put some sand over the spot so it will not stick to your shoes if you walk over it in the dark. Waterproof sealers protect driveways from the elements.

BASEMENTS

Eliminating dampness and fixing wall leaks are among the most common basement repairs. Cracks in walls, mortar joints, or in the floor are the culprits. Any crack that continues to get larger should be examined by an expert. Stable cracks should be fixed as soon as possible, preferably when the weather is dry. They should be fixed before freezing weather, since the ice may just enlarge the damage. Any crack bigger than ⅛-inch should be chiseled to a wedge shape, letting the hole widen at the rear so it will take the patching repair material.

Cleaning out the crack after it is widened to remove even all the dust is important. Although there are concrete mixes that claim to be excellent for filling cracks, a two-part epoxy mix is superior. The adhesive costs more, but it works much better. The epoxy can be troweled into the crack. After about a half-hour, the repair line can be smoothed with just a moistened finger. Epoxy is especially good for small areas.

If inside-the-basement measures do not stop water problems, then you have to go to the exterior. Various exterior remedies will stop water from entering the wall. The best grades with the strongest adhesive and waterproof qualities will solve the situation permanently.

Unless a foundation is absolutely watertight, there are times when water may come up through the joint where the wall meets the floor. A crease should be cut out where the wall and floor join with a chisel. Remove all the debris with a brush and then vacuum the area. Put a two-part epoxy patching mix into the cavity, filling the area completely. A spoon is a good tool to smooth the filling for a neat seam. Wetting the spoon will make it easy, and the epoxy cures in 24 hours.

Chapter 6

Auto Repairs and Adhesives

Some years ago, a firm got some attention from the car publications when someone came up with the idea of putting up an adhesive against welding. Two tractor blocks with cracks were fixed, one with a welder, the other with glue. The latter was as strong or stronger than the welding. These heavy metal-to-metal operations with adhesives have been restricted primarily to industry rather than automotive uses to date.

EXTERIOR TRIM

Rocker panel molding, side trim kits, and other self-adhesive decorations are available for most cars, trucks, vans, and recreational vehicles. An array of trim is available for autos from the accessory manufacturers. Cars may be dressed up instantly with the moldings and trims that come ready to stick on by just peeling off the backing paper. Pinstriping may be

attached the same way with the pressure-sensitive glues already applied. The already applied adhesives on trims and moldings allow for custom creativity in decorating autos. These accessories also allow you to cover up missing paint or other body problems.

ADHESIVE COATINGS

There are spray-on undercoatings and sound deadeners that protect against rust and reduce the noise level inside the car. These special coatings dry within 20 to 30 minutes, and stick so well they are protective against salt and rust corrosion. Like any adhesive, there is a curing time for these car coatings, but this usually amounts to no more than a couple of days.

Just how many new cars leak around the windshield, windows, or trunk lids would be a most interesting statistic! Paying up to Rolls-

Royce prices for cars that leak for want of a tube of sealer is a sad comment on assembly-line production. Some of the most expensive luxury autos leak from time to time right off the showroom floor. This poor work guarantees the sealer-adhesive manufacturers of very steady sales. The finest automotive glass sealers remain permanently elastic, and they are harmless to paint and rubber. There are some combination caulks that are suitable for minor body repair as well as windshields.

There are some interesting and efficient gas tank sealers to stop leaks. Some of them will permanently repair rusted tanks, according to the manufacturers. Some of these remedies are especially good for fixing antique gas tanks. In many cases, replacement tanks are not available, and the reproductions are often very expensive. To use some of the best ones, the gas tank must be removed from the vehicle and all the parts—float or other internal workings—taken out, as the sealer might clog anything left inside the tank. The sealer is poured into the tank and sloshed around.

After the inside of the tank has been thoroughly covered, it should be allowed to air-dry for one to three days before putting it back in the vehicle. One quart, which retails for $10 to $15, will seal most tanks. There are also metal tank leak-stoppers that can be applied to the outside of the tank. Sealing bad spots inside *and* outside is advisable.

Automatic transmission sealers form a pliable coating to build up cracked or worn seals and often stop leaks. They seal almost instantly and will also take care of minor housing cracks and line leaks. Some of these sealers eliminate engine racing due to slipping clutches or lazy shifting of the transmission. It is advisable to read the labels for the ingredients on all automotive sealers of any type. Quite often you can save money by buying the same formula for less money. The better-known—at least through advertising—brand names are often more expensive than cans sitting right next to them on the shelves that contain the same stuff for less money.

Power steering units must do a lot of work. Sealers for many automotive uses years ago were primarily seal swellers—they just bloated up an already damaged gasket. The best sealers today have adhesive qualities as well as other properties for taking care of particular problems and parts. A power steering sealer can save on costly repairs by sealing cracked or shrunken seals. The better ones will also stop cracked housing leaks and eliminate squealing sounds in the unit. They seal those hard-to-get-at leaks without disassembling any parts. These sealers may also be used for preventive maintenance, according to the directions on the products. For less than $5 some of these sealers might solve problems that would cost hundreds of dollars to repair in a garage.

RADIATORS

Leaking radiators are expensive to fix in auto repair shops. Just taking off some car radiators to repair them can be a mammoth and costly job. Hermetite's Leakstopper is one of the best to stop radiator leaks, but it must be applied to the *outside* of the radiator. The compound is not one of those that is poured in like a liquid sealer. Leakstopper is a heavy, thick material that even works under water. It is not affected by gas, oil, alcohol, or many chemicals.

UPHOLSTERY

Adhesives attach a lot of auto upholstery. Upholsterers are using more and more glue to weld seams instead of sewing them, even in the furniture world. Headliners are glued, door panels are glued, and many pieces of trim are glued. There are many kinds of upholstery materials on the market that you can buy to completely renew an automobile.

There are patching kits for matching different kinds and colors of upholstery, and there are repair tapes for upholstery that will fix rips in vinyl or leather. They will cover spots or prevent further damage. The good pressure-sensitive tapes will last as long as the original upholstery. Headliner trim kits are not usually already coated with an adhesive; they just include a tube since each job differs in application.

One of the worst glue jobs found in most cars is in the trunk, where the carpet has come loose. Most car manufacturers just tack the carpet around the ends, and it often comes loose, wrinkling up and causing problems. One of the most sensibile solutions to this is to use one of the carpet adhesives that will allow you to lift up the carpet at any time—it just peels away from the glue, but it reseals itself easily.

One of the most powerful glues used in auto repairs is that in which the plugs are dipped in fixing flat tires. The tire repair people just dip the plug into the glue, holding it with the special tool that fixes flats. It bonds instantly and cures almost as quickly. It is rather rare for a plug to come loose if it has been properly installed.

Van conversion kits that include upholstery and trim are very dependent on glues. These should be checked carefully to ensure that they are of high quality. Age affects some of them in particular. There are some good vinyl repair kits that will solve damage to car tops, seats, door panels, dash padding, van furniture, or literally anything else made of vinyl, leather, or Naugahyde. They will mend cuts, holes, cigarette burns, scuffs, torn seams, and more. They will restore upholstery to its original color, strength, and texture. Color and even surface grain can be matched quickly. The liquid vinyl may also be used to make very quick emergency repairs when color matching isn't important. In addition to the adhesive, you generally need a piece of the material that is going to be repaired. Often, this scrap may be taken from an area underneath the piece.

BODY REPAIRS

Fillers and adhesives to hold them in place sum up a lot of auto body repairs. There are many body repair compounds; shop carefully to get the right product for the job to be done. With holes of various sizes, some sort of backing usually has to be used, generally screening, to hold the filler in place. The area must be prepared well before the patching begins. In some instances, small holes must be drilled to serve as anchors for the materials. Professional body repairmen fix severely damaged autos to perfection. Even experts can't readily detect that they have ever been damaged. Doing as neat a job as possible is important because the excess must be sanded away to get the perfect surface to accept the paint.

FIBERGLASS

Watertight fiberglass patches can be done when you are faced with a jagged hole in metal, plastic, or even wood. The same work and materials—although the adhesives may differ—apply to fixing a plastic car body to a plastic laundry tub. Polyester and epoxy resins are used with fiberglass. Many prefer to use the polyester because of its low cost, workability, and short drying time compared to the thicker epoxy resins, even though they bond more readily.

Fiberglass repairing and laminating employ the same basic steps, overlapping layers of resin-laced cloth over a damaged spot. Getting the layers down precisely and obtaining the right resin mixture are essential. Mixing liquid plastic resins must be done with precision, taking care that the ingredients are not subjected to any moisture or air bubbles. An accurate scale should be used to measure out the weight of the ingredients for epoxy and acrylic resins.

The drops of hardener for polyester resins must be countered carefully, since the proportion is totally dependent on what properties are going to be needed—the more hardener used, the shorter the drying time. Humidity, temperature, and the thickness of the mixture determine the length of the curing time. Mixing some test batches and examining them for the results desired is the smartest way to get the most appropriate patching material. Inspect each mixture and use it under as similar conditions as possible during the testing. If it cures too fast, it will be too brittle; if it cures to slowly, it will be too rubbery.

To prevent air bubbles from getting into the finished resin, handle it carefully, and pour the ingredients slowly down the side of the container. Stir gently but thoroughly, making any bubbles rise to the surface. Running the paddle too quickly by hand or with a power tool will cause bubble problems; mix as slowly as possible. The mixing containers should be clean. Avoid any with wax linings, since they could contaminate the resin. A kitchen spatula or flat stirring stick works well.

The resins and hardeners must be weighed separately, and the sequence calls for pouring the hardener into the resin. One drop at a time on the directions means it. For resin and putty preparations, pour the polyester resin into the polyester putty, adding the cream hardener as directed. One way to measure is to squeeze certain lengths of hardener on the stirring blade.

Acrylic and epoxy resins can be colored by adding pigments to each ingredient a little bit at a time until the desired hue is obtained. To color polyester resin, add a little pigment to the resin only before mixing with the hardener. Resin and putty combinations are mixed first, and then the pigment is inserted, followed by the hardener. Fillers may be added, such as powdered stone, wood, or metal, just like pigments. Epoxies should have the resin first with a diluent; then add the hardener, followed by the filler.

Most of the time, three percent hardener by weight is added to resins. Curing may be speeded up by warming the area to be repaired with a heat lamp or hair dryer. Below 54 degrees F, cross-hardening will enable a cure down to 32 degrees. This technique involves mixing the standard hardener with the resin with the addition of a second hardener containing benzoyl peroxide. Directions must be followed specifically.

The spot that is going to be patched must be clean and free of paint. If the object being repaired is metal, it should be shined with a wire brush, washed with soapy water, alcohol, or vinegar, and allowed to dry. The metal must be degreased. The spot should be primed with a plastic filling compound.

Fiberglass is slippery. Since it is so difficult to handle, it is easier to soak the fiberglass in resin and then carry it to the repair spot on a sheet of plastic. This gives the fiberglass support so it will not fall apart, and keeps down the dripping. (Some fiberglass authorities declare that women are much better at laying up fiberglass than men; they seem to have an innate touch, making it less of a challenge with which to work.) Getting up fiberglass is difficult, so the work area should be kept as clean as possible and protected with a carpet of papers to catch drips. Once it cures, the task is formidable. Using throwaway containers is best to preclude any cleaning.

Fiberglass work is very dangerous, since the resin and the hardener are caustic, toxic, and flammable. They can be deadly to the skin and eyes so protective clothing and goggles are needed. To keep the fiberglass out of the lungs, a respirator is a must. This is particularly true even when sanding or cutting fiberglass. The air gets filled with the glass particles, which are easily inhaled.

A plastic auto body with a hole in it should be shaped with a saw to make an easier-to-work-with shape. Sanding with 80-grit paper on an electric drill or sander, feather a two-inch or so bevel around the hole. Using some cardboard wrapped in plastic, make a plate backing and tape it securely to the backside of the hole. Make sure the plate's plastic is facing the hole to be filled.

Scissor out a piece of fiberglass that will cover the hole and the feathered area. Next, cut a fiberglass mat just slightly smaller than the first piece. Continue cutting pieces smaller and smaller for a pyramid effect until the pile is the same thickness as the place to be patched. The last piece of fiberglass should be at least an inch bigger all around than the hole. Fraying the ends will help get the piece as flat as possible.

Put down a sheet of plastic on which to work, laying down the largest piece of fiberglass first. After the resin and hardener are mixed, take a soft-bristled brush and saturate the cloth with resin. Dabbing it on will help prevent the cloth from wrinkling. Continue to stack up the alternating layers of glass cloth and mat until the proper thickness is reached. Lifting the plastic sheet, center the fiberglass over the hole and gently press it in place.

To get rid of the air bubbles, leave the plastic wrap in place, and use a plastic squeegee. The bubbles will appear as white spots; work them out from the center to the edges. After the resin cures for several minutes, gently peel away the plastic wrap. Within two to five hours the patch will cure, and the cardboard backing may be removed. An electric sander or drill with a sanding disc will allow you to make a shallow depression in the patch. Feather it out for two inches beyond the original hole. Fill the depression with pieces of glass mat ranging from the size of the original hole to one one-inch out from it on all sides. Then insert two pieces of glass cloth the size of the feathered area around the hole, filling up the depression.

Mask off the area so the final inside patch may be applied without getting the undamaged spots dirty. With a new batch of resin, paint the repaired section before putting on the smallest piece of mat. Keep building up with pieces of mat until the last one is just a little above the surrounding surface. Work out the air bubbles with a squeegee as before, making sure you have a plastic sheet over the work area. When the patch hardens, gently peel the plastic away. A putty knife will cut away any loose edges or strands. Don't disturb the patch!

The gel coat comes next, applied with a soft brush. Put a plastic sheet over the coating so you can squeegee out the air bubbles. After it hardens, remove the plastic wrap. Using a fine grit wet-or-dry paper covered with fiberglass rubbing compound, sand the area very lightly. Brush on a second gel coat in the same manner as the first. After it cures, polish to match the rest of the surface with a buffing wheel and rubbing compound.

Working with large amounts of resin dictates that repairs should be done outdoors or in a well-ventilated place. The resin cures too slowly when the temperature gets below 65 degrees F, and hot or cold drafts should be avoided as well as direct sunlight. The best temperature range is 65 to 83 degrees F.

FIBERGLASS FLAWS

There is a sea of bad resin out there, and many poor fiberglass products. Some companies do a very lousy job. Errors in measuring and mixing are all too common—too much or too little hardener, the resin is too old, there was not adequate drying time, and wrong environmental conditions are among the problems.

Resin is dangerous. If a large quantity of resin is left open for around 45 minutes, it may start smoking, becoming a very real fire hazard! Activated resin generates heat as it cures—a lot of heat.

Common surface flaws and their causes include:

Crazing

Hairline cracks can appear immediately or several months later. They are due to incorrect proportions in the resin mixture, or the laminating resin was not compatible with the gel coat.

Star Cracking

If the gel coat has been applied too thickly, cracks may radiate from the area. A sharp blow to the underside of the area will also cause such cracking.

Fiber Pattern

A raised pattern effect is usually the result of failing to put a layer of the surface mat fiberglass between the gel coat and the woven roving. If the pattern is visible but smooth, the gel coat was too thin. If it was made in a molded form where the gel coat is applied first, the gel coat was not stiff enough when the fabric was applied.

Pinholing

When tiny air bubbles get into the gel coat, the surface always pocks. These bubbles generally were not removed during mixing because thay are difficult to see, especially if the gel is colored. The bubbles may be visible when the gel coat is too viscous, but they are hard to remove.

Leaching

When the resin washes away—it was improperly cured or the resin was not waterproof—the fiberglass fabric is exposed. The glass is fragile, and the patched area is weakened substantially. Most flaws can be fixed with a second gel coat over the surface, but leaching is too serious. The flawed area must again be cut away and replaced.

Chapter 7

The Sea of Adhesives
for Boat and Marine Repairs

Marine adhesives help keep the boating world and commercial shipping afloat.

The following article is reprinted with the permission of *Motor Boating & Sailing*. It was edited by Bernard Gladstone.

If you've tried to buy an adhesive lately, you know how difficult it is to sort through the enormous variety of products on the market. According to many of the labels and the advertisements, each sounds as if it could stick anything to anything under any conditions. While it is true that some can be used for several purposes, there is no such thing as a universal adhesive. For a permanent and satisfactory job, you have to match the adhesive to the task at hand—and that means knowing something about the kinds of adhesives suitable for the marine environment.

WOODWORKING GLUES

The three most useful wood glues are white polyvinyl acetate glues, plastic resin glues, and resorcinol glue.

White glues, the most popular of all for general woodworking, have moderate strength but comparatively poor resistance to moisture and exposure to dampness. For this reason, these glues generally have no place around the inside or outside of a boat. Even if the joint is located where it will never be exposed to water, dampness can still be a problem.

Plastic resin glues are powders you mix with water. They form a very strong joint—one that is often stronger than the wood itself—and they are very water-resistant. They are not completely waterproof, however, so they should not be used in places where the joint will be constantly underwater. They work best in joints that are reasonably snug-fitting and without voids, and they should only be used when working temperatures are above 65 degrees. The glue dries dark so it leaves a visible glue line in the joint, and it must be clamped for at least 8 to 12 hours to ensure a good bond.

Resorcinol glue is a two-part adhesive that consists of a powder and a liquid catalyst that you mix together. It forms an exceptionally strong bond and is completely waterproof, so it can withstand constant immersion. It is, therefore, the preferred glue for assembling or repairing wood joints below the waterline, as well as those that will be constantly exposed to the weather or submerged in bilgewater. Resorcinol is fairly good at filling gaps and voids in poor fitting joints, and like the plastic resin glues, it requires overnight clamping to ensure a permanent bond.

EPOXY ADHESIVES

These are probably the most versatile of all adhesives for use around the boat. They are two-part compounds that consist of a resin and a catalyst, and once mixed will bond to wood, metal, fiberglass, and most plastics. For a good bond the surfaces to be joined must be absolutely clean before the adhesive is applied, but once cured the material is impervious to attack by water, oil, gasoline, and most of the cleansing compounds and solvents likely to be found around the boat.

Expoxy adhesives come in all kinds of packages—tubes, aerosol cans, and jars—and in different types of formulations. Some are designed for mixing in equal quantities—resin and catalyst—while others are designed to be mixed in different proportions. Also, they vary as to curing time—some harden in as little as five minutes while others take as long as 8 to 12 hours. As a rule, curing time is affected by temperature; the colder it is, the longer it takes to cure.

Epoxies harden by chemical action. They don't need air to dry, so most will even harden when underwater once they have been mixed.

They are excellent for filling voids and gaps. No clamping is needed, as long as you bring parts in firm contact with each other in proper alignment.

SILICONE RUBBER ADHESIVES

Although most boat owners think of silicone rubber as a caulking and sealing material rather than an adhesive, this compound is also useful as a bonding material where considerable flexibility is required in the joint, along with moderate resistance to stress. It bonds well to glass, metal, wood, and fiberglass, and remains rubbery over a wide range of temperature extremes—from 60 degrees below zero to 450 degrees above. Thus, you could use this inside an ice chest, or on the outside of an engine block where it will be exposed to considerable heat while the engine is running.

INSTANT OR ONE-DROP ADHESIVES

These are the "miracle" cyanoacrylate glues that come in very small tubes, one to three ounces, because on most jobs only one drop is needed to make the bond. They stick within seconds and will bond to almost anything, including human skin, so they must be handled with caution. They work best, however, on nonporous materials such as glass, metal, and many hard plastics, and they must be applied in a thin film to cure properly.

Because these adhesives dry so rapidly, no clamping is required—just press the coated parts together and hold them for a few seconds. But the thin film means that the parts must fit together snugly without gaps or voids, and the quick-setting characteristic of this adhesive also means that you have to align parts accurately the first time you bring them together—shifting or moving pieces around will often ruin the bond that is starting to form.

The adhesive forms a joint that is almost as strong as one formed with an epoxy when the parts fit properly, so these quick-drying adhesives can be a handy addition to any boat owner's emergency repair kit. Although claims differ, it is generally agreed that the joint formed by a cyanoacrylate will not be quite as waterproof as one formed by an epoxy.

TWO-PART INSTANT ADHESIVES

This is the newest of the "miracle"adhesives. It forms a very quick bond that is also very strong and highly water-resistant. Falling somewhere between cyanoacrylates and epoxies in strength, it sets almost as fast as the "instants" described above but gives a bit more time to align parts and bring them together. You apply one part to one surface and the second part to the other surface . . . but the actual cure doesn't start until you bring the two together. Hold for a minute or two, and the bond is made . . .

Unlike the cyanoacrylates, it will bond well to flexible materials and many porous surfaces, so it does not require a perfect fit between mating parts. Also, it has one very unusual quality that can be of great value to boat owners—the surfaces to be joined do not have to be absolutely clean. As long as loose dirt is removed, it will work on surfaces that are oily, greasy, and grimy. Just brush the liquid activator on the surface, squeeze the acrylic adhesive onto the other surface, then bring the two together. You only have to hold the parts together for a minute or two.

CONTACT CEMENT

Made with a synthetic rubber base, these cements are designed to do just what their name implies—bond on contact. They are used for such things as putting plastic laminates on counter tops, furniture, cabinets, bulkheads, and similar surfaces because they eliminate the need for clamping or applying weights while the adhesive sets. The bond formed has a moderate amount of strength that is adequate for laminating and similar applications where very little actual stress will be encountered, and they can be used to join metal, rubber, leather, and many plastics to wood or to each other.

Unlike most other adhesives, contact cement must be applied to both surfaces, and it must be allowed to dry before the parts are assembled. A surface coated with contact cement will only adhere to another surface coated with the same cement. But once the two parts are brought together, after both sides are dry, they bond instantly with a tenacious grip that allows for no shifting or realigning of pieces. You have to get them right the first time.

When large sheets of plastic are being laminated to a surface, use a "slip sheet" of brown wrapping paper. Place the wrapping paper between the sheet of plastic and the surface to which it will be cemented, then line the two pieces up carefully—the cement won't stick to the paper because it has no cement on it. When the pieces are in position, you lift one corner of the plastic, then grab the paper and slide it out carefully. As the coated surfaces come together, they will bond instantly.

VINYL REPAIR CEMENT

Designed specifically for repairing tears, rips, and punctures in vinyl upholstery and vinyl-covered cushions, these adhesives are quick-drying cements that set up in a few minutes and cure completely in a few hours or less. They contain solvents that virtually "weld" the torn vinyl together, and are handy to have on board for repairing weather curtains, cockpit covers, and vinyl-covered furniture. The cement dries clear so in most cases repairs are inconspicuous, but one company also packages this cement with a small touch-up kit to color the patch so that it becomes completely invisible.

One of the most difficult areas of application on boats is in the forward compartment deck sections or the cabin roof lining. Lots of liner material is sagging on boats. Heat from the sun has broken the bond of an inferior adhesive. Material sag or separation will not happen with the best marine adhesive.

Sideliner materials including decorative trim, marine carpeting, vinyl-coated fabrics, quilted materials, and integral skin foams can be permanently bonded to interior hull surfaces and cabin interior surfaces with a spray gun, roller, or brush application.

Marine adhesives may be used to bond sound-deadening, heat, and flame-resistant materials to engine compartments and in galley areas such as above stoves and refrigerators. Every boat should have a stock of every possible type of glue in its supplies. Repairing furniture, utensils, and anything onboard may be important when on an outing. Repairs to the interior or exterior of the boat may be possible immediately if the proper glues and adhesives are available.

Table 7-1 details the various glues and adhesives useful in marine applications.

BOATBUILDING

Boating is sailing into society and lifestyles with tremendous growth. Boat sales projections are extremely high, with millions of people planning on owning a boat. More and more people are discovering that they can build their own boats and save approximately half by doing so.

Clark Craft Boat Co. of 16 Aqua Lane in Tonawanda, New York 14150, is one of the finest sources for the do-it-yourself boat builder. The firm offers four ways to build your own boat.

Plans and Patterns

Those who are well-informed and experienced in the boating world may prefer to build their own boat from plans and patterns. They must have all the necessary tools and woodworking machinery and must furnish all the materials. Clark Craft declares it has the largest selection of quality plans and patterns available anywhere. These large-scale plans are easy to read, and each set has blueprint drawings and detailed instructions.

Frame Kits

Clark Craft's complete frame kits are completely cut out and notched. Frame kits and frame sets provide the do-it-yourselfer with the most economical way to build boats and be assured of the frames turning out true to form. All difficult parts that require power tools and factory jigging have been precut and preassembled using marine glues and silicone bronze anchor fast nails or screws. The frames are made of top quality solid mahogany airdried to Clark Craft's specifications. Each frame kit includes a complete set of frames, blueprints, building instructions, and patterns for the additional parts needed to complete the boat ordered.

Frame sets are available for most designs for those who do not want to spend the additional time fabricating their own frames. You should be familiar with frame line drawings if you want to put your own frame kit together.

Hull Kits

Clark Craft may be the only firm in the industry that offers complete hull kits. These kits include all the parts to completely build

Table 7-1. Marine Adhesive Guide.

Adhesive	Mixing Required	Clamping Required	Drying Times	Water Resistance	Bond Flexibility	Materials Bonded	Special Characteristics
Resorcinol glue	Yes	Yes	12-18 hours	Excellent	Not Flexible	Wood, Leather Cork, Textiles	Will fill gaps in poor fitting joints.
Plastic Resin Glue	Yes	Yes	8-12 hours	Good	Not Flexible	Wood	Won't fill gaps and requires snug-fitting joints, requires temperature of at least 70 degrees.
Epoxy Adhesives	Yes	No	1-8 hours	Excellent	Not Flexible	Metal, Fiberglass, Wood, Glass, Ceramics, some Plastics	Low temperature slows drying and curing, has excellent gap-filling ability, will harden under water.
Silicone Rubber Adhesives	No	No	1-2 hours	Excellent	Very Flexible	Metal, Painted Wood, Glass, Fiberglass, Rubber and Plastic	Resists extremes of high and low temperature and stays flexible.
Contact Cement	No	No	Instant (after contact)	Fair to Good	Moderately Flexible	Wood, Plastic, Leather, Rubber Fabric, Vinyl	Both surfaces must be coated and allowed to dry beforehand.
"Instant" or "One-Drop" Adhesives	No	No	30-90 seconds	Very Good	Not Flexible	Metal, Hard Plastic, Glass (some on Wood)	Works best on nonporous surfaces. Must be used in thin film, won't fill gaps.
Vinyl Repair Cement	No	Yes	30-60 minutes	Excellent	Very Flexible	Vinyl and Leather	Clamping consists merely of using masking tape.
Two-part Acrylic "Instant" Adhesive	No	Yes	2-4 minutes	Excellent	Moderately Flexible	Metal, Wood, Fiberglass, Ceramics, Glass some Plastics	One part is applied to one surface, the other part to the second. Curing starts when they come together. Can be used over oily and greasy surfaces.

the hull ready for the water, minus the paint and hardware. All plywood, sides, and bottoms are precut and ready for assembly. All frames, the transom, the stem, and other parts are preassembled and ready for installation.

Boat Kits

Today's boat kit is a precision package. Each piece is cut and assembled in jigs to fit perfectly. Kits are available for plywood, ferro cement, and the new foam fiberglass construction.

WOOD AND EPOXY BUILDING SYSTEM

"With more than 40 years of boatbuilding experience, and being a pioneer in the wood and epoxy boatbuilding and repairing techniques, we now offer the most complete selection of quality equal mix epoxies available," Clark Craft points out.

Clark Craft tested literally hundreds of epoxy formulas from various companies before having the Isochem Corporation produce a compatible equal mix system, a low viscosity wood saturating resin, epoxy glues, and epoxy surfacing compounds, all of the highest quality for marine uses.

Armor Epoxy

This marine epoxy is an equal mix 100 percent epoxy coating applied with a brush or roller in two separate coats. The first coat soaks deeply into the grain, creating a barrier or locking coat. The second coat leaves a smooth finish and creates a tough, permanent coating against any future vapor penetration. Besides adding tremendous strength and dimensional stability, Armor Epoxy will preserve the wood against any future dry rot. It has been designed to stay extremely flexible and will not dry out or become brittle over the years. If additional abrasion resistance is re-

quired on high-speed powerboats, Clark Craft recommends using a silane treated cloth over the exterior of the hull.

GL 10 Super Epoxy Glue

Clark Craft says its Super Epoxy Glue is the best of its kind. It has a thick yet spreadable consistency so that it will easily stay in place when spread along a horizontal surface. Under high-stress conditions, such as an impact, the glue will hold fast, and the wood fiber parts on either side of the glue line will generally shatter, the mark of an excellent glue. It bonds wood, fiberglass, concrete, glass, and most other materials. It is an equal mix formula, has no shrinkage, cures at 35 degrees F, cures on moist wood, produces stress-free joints, and applies with contact pressure.

Epoxy Surfacing Compound

Clark Craft's GL 15 equal mix surfacing compound is excellent for boatbuilding or automotive body repair. Formulated to a smooth, creamy consistency, it may be used for surfacing large hull areas for small imperfections. It is ideal for filling screw or nail holes and may be sanded to a feather edge.

Armor Poxy Wood Saturation System

In order to determine the proper amount of Armor Poxy resin to order for a boat project, you have to decide if you are going to use the two-coat system, which is highly recommended for full protection. If you were going to build a small kayak, you would probably use AB exterior grade plywood. To keep the cost as low as possible, you could get by with only one heavy coat of Armor Poxy. While one coat may be acceptable for small cartop boats, it would not be appropriate for maximum protection of larger boats.

The Work Area

If you build a boat outdoors, do not work in direct sunlight; if the building is done indoors, be sure there is good ventilation even though the toxicity of Armor Poxy is extremely low. For the best wood penetration, the temperatures range should be 70 to 85 degrees F. When constructing a boat under cooler conditions, electric heaters and large heating bulbs should be used to heat the hull or deck prior to application. Be sure the Armor Poxy itself is also at least 70 degrees. If the container has been sitting on a cold garage floor for hours, it may be too cold to apply. Leaving the Armor Poxy in a warm area until it is ready to use may be advisable when working in cooler temperatures.

Mixing Armor Poxy

The mixing proportions should be by volume only. One pint of resin should be mixed with one pint of hardener. At 70 to 75 degrees F, the pot life of the glue should be 30 to 40 minutes. Clark Craft notes the best way to mix the Armor Poxy is to use a stick about an inch wide to stir the batch for about two minutes, or some 150 to 200 strokes. Stirring it with an electric motor is not recommended because this tends to whip air into the epoxy, which is very difficult to remove later.

Temperature is extremely important! Although 75 degrees may give you 30 to 40 minutes pot life, a temperature of 80 to 90 degrees will shorten your pot life to just 15 minutes! The coated surface should be set up in four to five hours at 70 to 75 degrees. The second coat may be applied directly over the first without any sanding. If there are any runs on the vertical surfaces, they should be sanded smooth before applying the second coat over them.

If the temperature is 85 to 95 degrees, the working life of the epoxy is extremely short. You can extend the pot life in this hot weather by working with smaller batches of the glue. It will help if the mixture is poured into a shallow pan, such as a pie tin, where the depth of the mixture does not exceed one-half inch. This will help dissipate the heat greatly and may double the number of minutes of pot life. Packing some ice around the container it is in will help reduce the heat and increase the pot life also.

Coating the Parts

The first saturation coat should be applied with a brush or roller. It will be similar to painting with latex paint. Keep applying the coating until it has a good wet look. Dry wood will absorb the coating so that an area that looks like it is well covered may begin to dry out in several minutes. This is especially true of end grain areas, such as the edges of frames, and especially plywood paneling. The first coat is critical, so you must supply the wood with all the resin it can absorb. If you can determine ahead of time specific areas on the frames and especially on the stem, chine, and keel where further work will have to be done, such as planing, cutting, and sanding, these areas should be left bare. If they are coated accidentally, this is no problem since they may be recoated after they have been faired.

Coating Plywood Parts

If the building project is a boat and the framework has been erected, the plywood side and bottom panels should be fitted together, but do not fasten them permanently. These panels should be precoated before installation, and the best way to do this is to lay them down on the ground or bench so they are in a stationary horizontal position. The first coat of Armor Poxy is easily applied then, preferably with a paint roller. Use the epoxy very liberally so the entire plywood panel looks wet when the coating is completed. Some dull areas will appear after a few minutes because the softer grain absorbs the epoxy more quickly. Recoat these areas immediately, while they are still wet. Stroking the plywood panel with a three or four-inch brush will level the stipple effect laid down by the roller, giving a fairly smooth surface. This must be done while the coating is still wet.

After the side is dry, flip it over and repeat the coating process on the other side. After both sides have been coated and dried, select the side that is going to end up on the inside of the boat. Now apply a second coat of epoxy the same way the first was applied.

When you use the paint brush to stroke out the coating, it will level just like a sheet of glass, and there will be no dry areas. Depending on the temperature, the epoxy should become tacky in about one hour. The panel is then ready to be fitted directly in place and fastened to the precoated battens, keels, chines, sheers, or wherever the panel is to be installed. When the panel is installed in this way, it bonds directly to the batten strips and may never be removed! The panel has already been coated on the inside, saving many hours of trying to paint all those little nooks and crannies.

Armor Poxy as Super Glue

Using Armor Poxy as a super glue produces structurally perfect joints. Thickened Poxy should be used when gluing precoated frames or plywood paneling in place. Adding a small amount of thickening agent, the resin can be varied in thickness considerably. It is an unchallenged structural adhesive. Clark Craft advises not to use resorcinol or any other type glue for this purpose. The desired thickness depends on how much of a gap the epoxy is required to fill, and it is important that the resin fill all gaps in the joint for maximum strength.

Final Finishing

It takes about two days at 70 degrees F for the Armor Poxy to harden before it is ready for painting. A waxy film may appear on the surface after it hardens, and this may be removed by washing it with a solvent or detergent and water.

Armor Poxy resin should be kept off the hands and clothing as much as possible during its use. Any epoxy resin or hardener on the skin should be removed with a sudsy household ammonia. Do not use solvents such as acetone or lacquer thinner because they tend to dry out the pores of the skin and increase irritation. People allergic to the resin should wear inexpensive polyethelyne gloves.

Armor Poxy has an almost indefinite shelf live. Clark Craft has had some samples being used in its shops for more than five years. This is quite an asset since it has many uses around the home. Any wood surface can be made waterproof, alcohol, and acid resistant by coating it with the special epoxy. It is an excellent coating for shower stalls, fixing leaky pipes, water spouts, and gutters. Outside wood steps can be preserved and protected with the epoxy, and it is great for damp basements and cellars. Two or three coats must be painted over dry cinderblocks or cement walls to create a positive and lasting moisture barrier.

Clark Craft has an ideal beginner's boat in its PB-1, which it calls the "Punkin Seed." The sailing dinghy complete kit, including the sail kit and Dacron sail, sells for $399 F.O.B. The pram complete boat kit is just $199 F.O.B. (Figs. 7-1, 7-2).

Fig. 7-1. This trim pram is an ideal boy's boat. Courtesy Clark Craft Co.

Primarily designed as a boy's boat, this useful little sailing pram can be used as a tender for a larger cruiser, is easy to row, and under power performs well with up to a 5-hp motor. The sailing dinghy version makes an excellent safe sailer for the beginner and is almost impossible to capsize due to its minimum frames along with mahogany longitudinal members, such as keels, chines, sheer, and other parts. It is all precut ¼-inch plywood bow and stern transom, reinforced with ¾-inch solid mahogany framing all waterproof material. Screws, glue, and instructions are all furnished. The optional sail kit includes an aluminum mast, mahogany boom, dagger board, rudder, the fittings, gooseneck, pulleys, masthead, and all other fittings to convert this little pram into a sailing dinghy.

Clark Craft urges the beginning boat builder to study the kit's instructions carefully. Separate drawings are furnished for each assembly stage. Parts to be installed are listed in order of assembly on each drawing.

Proceed through the step-by-step assembly procedures, selecting the corresponding parts from the packaged boat kit. You should

Fig. 7-2. Building the boat doesn't take much room. Courtesy Clark Craft Co.

know each part and where it goes and how it is installed. A building jig or platform to assemble the boat is not needed. Engineering and construction details have been carefully checked. Cutting or trimming in the entire assembly has been reduced to a minimum.

Boatbuilder, the magazine for boatbuilders and yacht designers, is a fine publication for anyone interested in building boats. The publisher and editor of the magazine, Keith E. Lawrence, is one of the world's foremost authorities on building any size or type of boat. The magazine's address is P.O. Box 1109 in West Palm Beach, Florida 33402.

Chapter 8

Glues and the Workshop

Articles appear periodically in the home workshop type magazines on how to glue various things. Some give information on glues and adhesives, but they are almost never done in sufficient depth. An article that appeared in *Fine Woodworking Magazine* written by George Mustoe is an outstanding exception. This very high quality magazine is the finest in the woodworking field. Mustoe is a geochemistry research technician at Western Washington University in Bellingham, Washington. The following article is reprinted with permission from *Fine Woodworking*.

Like the alchemists' attempts to transmute base metal into gold, much human effort has gone into the search for the perfect glue. This goal is probably as unrealistic as the dreams of alchemy, but the inventors' struggles have not been without reward: Adhesives manufacturing is a big growth industry in the United States, and per capita consumption is about 40 pounds per year.

Not surprisingly, "What kind of glue did you use?" is a frequent query heard whenever woodworkers gather. Unfortunately, these exchanges generate some old wives' tales, among them the colorful but incorrect assertion that cyanoacrylate "superglue" is derived from barnacles.

Because wood is a relatively weak construction material, most adhesives produce bonds that are stronger than the surrounding lumber, so claims of extremely high strength are seldom meaningful to the woodworker. Instead, the most important characteristics are setting rate, viscosity, resistance to water, flexibility, color, sandability, and gap-filling properties. As a woodworker who happens also to be a chemist, I've developed a keener-than-usual interest in the literally hundreds of glues sold today, discovering in the course of my research that only about a dozen kinds are useful for woodworking. Within each category, I've found that different brands will usually perform equally, so the choice for a particular project is best made by understanding the chemical makeup and characteristics of the glues we use.

In this article, I'll cover those glues that are best suited to general woodworking. In a second article I'll talk about epoxies, hot-melt glues, cyanoacrylates and contact cements, all specialty glues that are usually more expensive, though not always better, than our old standbys.

PROTEIN GLUES

The natural world abounds with examples of sophisticated adhesives which display impressive tenacity; barnacles and mussels, for example, cement themselves to beach rocks and ship bottoms with a substance that resists prolonged immersion in salt water. Though the chemistry of these natural adhesives is poorly understood, most sticky secretions are combinations of various complex proteins. Thus it is not surprising that early artisans discovered that the best raw materials for glue were protein-rich animal products such as skin, bone, and blood.

Today, despite the advent of modern synthetic adhesives, animal-protein glues are still common. They can be divided into three types: hide and bone glue, fish glue, and blood glue. Of the three, hide and bone glues are of the greatest interest to the woodworker. The use of fish glue, which is derived from the water-soluble proteins in fish skins, is limited to industry, mainly for attaching labels to bottles and occasionally as a tack-improving additive to white glue. Blood glues, once developed as water-resistant adhesives for early military aircraft, are made by dispersing beef or pig blood in water, with wood dust, lime, or sodium silicate added as thickening. They're

most often encountered in vintage plywood, but are practically impossible to buy today and have no significant advantages over readily available synthetics.

Hide and bone glues, on the other hand, are far from obsolete. Besides being widely used in industry for products such as gummed paper tape, sandpaper, and bookbindings, hide glue finds plenty of uses in the woodshop. The setting time and spreadability can be varied, and the adhesive cures into a colorless, non-toxic, sandable glueline which can be undone by the application of moist heat—a feature that is attractive to luthiers, for instance, who may need to remove the soundboard of an instrument to repair it. Water also softens hide glue, and some furniture conservators use a 50 percent vinegar solution to speed the disassembly and repair of antiques.

Hide glue consists of protein derived from collagen, the main ingredient of skin and connective tissue. The glue is prepared by cooking animal hides, hooves, and tendons into a protein-rich broth which is then cooled to a gelatinous solid, slicked, dried, and ground into a coarse powder. In retail stores, hide glue is commonly sold as a pre-mixed liquid, but it can be bought in powder form, in which case it must be mixed with hot water. Liquid hide glues have two advantages over mix-your-own: You don't need a heated glue pot, and the slow setting rate may be valuable for complicated assemblies. During the heyday of hide glue, it could be bought in 18 grades, each with a different viscosity and setting time. Today, woodcraft suppliers usually offer only a single, high-grade product. Setting time can be slowed by adding more water, but this leads to a slightly weaker bond.

Many other proteins have adhesive properties. Soybean-based glue is used in some interior plywood. Casein or milk glue, which has been detected in medieval picture frames, is made from skim milk, and is used today for laminating interior beams and trusses. This glue is a light-colored powder that must be mixed with cold water and allowed to stand about ten minutes before use. Unlike the other protein glues, casein sets both by evaporation and by chemical reaction, forming calcium caseinate. The resulting neutral-colored bond is highly moisture-resistant but not waterproof. Casein can be used in cool weather and on woods containing up to 15 percent moisture. It is particularly effective with oily woods such as teak, yew, and lemonwood. Powdered casein glue is available from National Casein, 601 W. 80th St., Chicago, IL 60620.

PETROCHEMICAL RESINS

Casein glue is sometimes confused with polyvinyl acetate (PVA) white glues. Part of this confusion stems from the milky appearance of white glue and also because dairy-related companies such as Borden, who once marketed casein, now sell PVA glue. Developed during the 1940s, PVA glue is part of a family of synthetic resin glues that have largely replaced animal glues in the woodshop.

Derived from petroleum compounds and acetylene gas, white glue consists of minute PVA globules suspended in water. When the glue is spread on wood surfaces, the water evaporates and/or diffuses through the surrounding porous material, and the globules coalesce to form a tough film. Because of its reputation as cheap hobby cement, white glue is sometimes undervalued as a woodworking adhesive. Actually, its low viscosity, rapid setting time, and fair gap-filling qualities make it an excellent choice for general woodworking. It dries into a clear, slightly flexible glueline, and it remains fresh on the shelf almost indefinitely. PVA is nontoxic, making it safe for use around children.

The major disadvantages of white glue are its low resistance to moisture and the gummy, thermoplastic nature of the dry film: It turns rubbery under the heat of sanding and clogs the sandpaper. You can minimize clogging by removing excess glue with a sponge or a damp cloth before the glue sets, or by trimming away hardened glue with a chisel or a scraper. The soft film also causes PVA-glued joints to "creep" out of their original alignment when subjected to continuous stress. While this may allow joints to adjust to seasonal variations in humidity without cracking, it's not a desirable quality if great structural strength is needed.

Be cautious when buying white glue. Competition among the 40 to 50 manufacturers of PVAs has kept the price low, but with the predictable advertising hype. In testing white glues, I found only one that yielded inferior results, a generic white craft glue distributed by a local hobby shop. Its adhesive properties compared favorably with leading brands, but the glue reacted with most woods to produce gray or black stains. Chemical analysis revealed that the glue was contaminated with high levels of dissolved iron.

In recent years, woodworkers have been attracted to another type of PVA adhesive, aliphatic resin or yellow glue. Actually, the label is a bit of a marketing ploy, since both yellow and white glues are technically aliphatics, which means that they consist of long chains of molecules. Yellow glues have qualities similar to those of white glues, but they contain polymers that speed tack time and improve moisture and creep resistance, at the expense of a slower final cure. Yellow glues are also less thermoplastic, so they won't gum up sandpaper as badly. . .

Yellow glue may be more difficult to apply because of its thick consistency, but it is also less likely to squeeze out when clamped. The viscosity increases as the glue ages in container. Manufacturers recommend that the glue be used within 6 to 12 months of purchase, but some workers successfully store it for up to two years by stirring in small amounts of water to reduce the viscosity. Up to about 5 percent water can be added without affecting bond strength. Freezing can ruin white and yellow glues, both in the bottle and as they cure. Manufacturers add compounds to improve freeze-resistance, but any PVA that seems curdled should be discarded.

WATER-RESISTANT GLUES

Modern industrial processes have been revolutionized by the development of highly water resistant synthetic resins, beginning in 1872 when the German chemist Adolph von Baeyer—of aspirin fame—discovered that he could produce a solid resin if he reacted phenal with formaldehyde. This basic chemistry forms the foundation of the plastics industry and has given birth to a family of versatile, reliable adhesives. Phenolic resins, because of their cost and heat-curing requirements, are used mostly in industry and for exterior plywood and water-resistant particle board. But a chemical cousin of the phenolic resins, urea-formaldehyde resin, is cheaper and easier to use, making it an adhesive of choice when water resistance is needed, or when long open time between spreading the glue and clamping up is desirable.

Phenolics and urea-formaldehydes cure not by evaporation, but by cross-linking or polymerizing their molecules into hard films that aren't softened by water. The small shop woodworker will be most familiar with the type that consists of a light brown powder which must be mixed with water before use.

Urea-formaldehydes are good general-purpose wood adhesives, especially for woods of relatively high moisture content. They cure into hard, brittle films which won't clog sandpaper, but, for the same reason, they are poor gap-fillers. The medium brown color when cured blends well with most cabinet woods, although bonding may be inhibited in some oily species, such as rosewood and teak. Most urea-formaldehydes aren't recommended for marine use, but they are sufficiently water-resistant to withstand sheltered outdoor applications.

When high strength is not essential, urea-formaldehyde can be extended by adding up to 60 percent wheat flour or fine wood dust. The thermosetting nature of urea-formaldehyde glues can be both boon and bane. In a shop cooler than 70 degrees F they will cure poorly or not at all, but at 90 degrees F the mixture's pot life is only one to two hours. Once the glue is spread and the pieces clamped, curing can be hastened by heating the glueline to between 90 degrees F and 150 degrees F. Urea-formaldehyde's thermosetting qualities make it the most popular adhesive for use with radio-frequency curing apparatus.

One drawback of urea-formaldehyde glues is the emission of formaldehyde gas during and after curing. Besides being a suspected carcinogen, this vapor may irritate the skin and eyes and cause headaches. The problem is liable to be most pronounced in homes constructed with urea-formaldehyde glued paneling, but it's a good idea to work with this adhesive only in a well-ventilated shop.

The development of urea-formaldehydes marked a milestone on the road to the perfect waterproof glue sought by boatbuilders for centuries. Ironically, completely waterproof adhesives didn't appear until the wooden ship was nearly extinct. Today, resorcinol-formaldehyde glue is the most popular waterproof adhesive, with epoxy resin trailing as an expensive second choice. Resorcinol glue was developed during World War II for gluing the plywood used in bombers, helicopter blades, and antimagnetic mine sweepers. Today, it is used to bond marine and exterior plywood, and for laminating outdoor timbers. For the home shop, resorcinol is sold retail as a two-part system consisting of a dark red liquid resin and a solid powder containing paraformaldehyde and an inert filler—usually powdered nutshells.

Resorcinol is fairly costly, and once mixed it must be used within an hour or two. For these reasons, it should be the glue of choice only when a completely waterproof joint is needed. It requires a minimum setting temperature of 70 degrees F, and solidifies within eight hours, though it doesn't reach full bond strength for several days. Acidic hardwoods such as oak may require 100 degrees F to 110 degrees F temperatures for maximum bonding. The final glue film is extremely durable, tolerating boiling water, caustic chemcials, and drastic temperature variations. Resorcinol glue is easy to apply and can be cleaned up with a damp rag. Disadvantages include the dark reddish glueline and the release of formaldehyde during curing. See Table 8-1.

When wood joints fall apart, as they occasionally do, the glue is automatically suspect. Usually, though, bond failure occurs not because the glue isn't strong enough, but because the wrong adhesive was used, the wood's moisture content was too high or too low, the surface was improperly prepared, or the joint was clamped incorrectly.

The wide range of glues available will meet any woodworker's requirements, but for most indoor woodworking, white and yellow glues are the best choices, except for veneering, where water-free glues such as epoxy or hot-melt sheets will keep the veneer from curling. Powdered resin glues can give erratic results due to sloppy mixing or poor temperature control, but they are excellent when a hard, machineable glueline is required, and for moisture-resistant exterior work.

Too much or too little moisture in the wood is one of the most frustrating causes of glue failure. Consider this example: The center of a two-inch thick board is liable to contain more moisture than the surface. If the lumber is planed and edge-glued before it reaches equilibrium moisture content, the porous end grain of the wood will dry and shrink faster than the middle, straining or breaking the glueline. To avoid this, stack and sticker your lumber after milling, postponing gluing until it has stabilized. An extra coat of finish on the end grain when your project is done will minimize subsequent stress on the glueline.

Climatic extremes can drive wood to equilibriums that will make gluing troublesome. In the desert Southwest, for instance, moisture content sometimes falls to 4 percent, which can draw the water out of PVAs before the joints can be assembled. Conversely, the glue won't harden in wood much wetter than 12 percent. In these environments, using adhesives that don't cure entirely by evaporation—urea-formaldehydes and casein glues—will help. Temperature can also be a factor in glue failure. Below 50 degrees F, PVAs come out of solution and cure in chalky weak gluelines. At high temperatures, say, about 100 degrees F, they are liable to skin over before assembly, which makes a strong bond virtually impossible.

Typically, adhesives bond to only the top layers of wood, so the surface must be smoothly cut, with no torn or partially detached fibers. Providing that it is straight and true, the best surface for edge-gluing is one left by a sharp hand plane. Next best is to use a jointer or even a sharp circular saw, preferably one that leaves indetectable sawmarks. Dull jointers and planers, on the other hand, produce a glazed, burnished surface which swells

Table 8-1. Conventional Woodworking Adhesives.

Adhesive	Application characteristics	Properties after curing	Recommended uses
Hide glue (hot)* (Behlen Ground Hide Glue, Behlen Pearl Hide Glue)	Fast track, viscous, min. curing temp. 60°F, moderate gap-filling ability, nontoxic, requires glue pot	Transparent, not water-resistant, can be sanded	Musical instruments, furniture
Hide glue (liquid)* (Franklin Liquid Hide Glue)	Slow-setting, low viscosity, min. curing temp. 70°F, moderate gap-filling ability, nontoxic, may have strong odor	Similar to hot hide glue	Assembly procedures that require slow setting
Casein glue* (National Casein #30, slow cure; National Casein #8580, fast cure)	Glue must stand 10 to 20 minutes after mixing, min. curing temp. 35°F, moderate gap-filling ability, nontoxic	Neutral opaque color, high water-resistance, sands cleanly	Interior structural applications, especially good with oily woods and in cool working temperatures
White glue* (Elmer's White Glue, Franklin Evertite, Weldbond, Wilhold R/C-56)	Cures rapidly, low viscosity, min. curing temp. 60°F, moderate gap-filling ability, nontoxic, almost unlimited shelf life	Transparent, low water-resistance, creeps under load, clogs sandpaper	General woodworking, not recommended for structural or outdoor applications
Yellow glue* (Elmer's Carpenter's Wood Glue, Franklin Titebond)	Fast tack, moderate viscosity increasing with age, min. curing temp. 60°F, moderate gap-filling ability, nontoxic	Nearly transparent, moderate water-resistance, less likely to creep under load than white glue, can be sanded	General woodworking, indoor use only
Urea-formaldehyde glue* (Weldwood Plastic Resin Glue, Wilhold Plastic Resin Glue)	Glue powder must be mixed with water, min. curing temp. 70°F**, poor gap-filling ability, releases formaldehyde vapor, uncured glue is toxic	Medium brown color, high water-resistance, sands cleanly, thick gluelines are brittle and may crack under stress	General woodworking, structural uses indoors or in sheltered outdoor locations, bonding may be inhibited with oily woods
Resorcinol glue (Elmer's Waterproof Glue, U.S. Plywood Resorcinol, Wilhold Resorcinol)	Moderate viscosity, min. curing tem. 70°F**, good gap-filling ability, releases formaldehyde during curing, two-part system must be mixed, uncured glue is toxic	Opaque reddish color, waterproof, withstands most solvents and caustic chemicals, can be sanded	Marine use and outdoor construction

* Water-based adhesive may cause warping of veneer or thin panels.

** May be rapidly heat-cured at 90°F to 150°F.

in contact with glue, encouraging failure. A sanded surface is similarly undesirable because the loose fibers left behind by the abrasive soak up glue but will part readily when the joint is stressed. Wood surfaces oxidize quickly, so try to mill and glue on the same day; machining a fresh surface on lumber that has been stacked for acclimation is advisable.

Mating surfaces should fit snugly without massive clamping pressure, but joints should have enough space to permit a glue film to develop; hammer-tight tenons or dowels will squeeze out most of the adhesive as they are assembled. If a joint is sloppy, don't rely on your glue's gap-filling qualities to rescue it. Better to recut the joint, or to salvage it with a strategically placed veneer shim.

Deciding how much glue to apply is a dilemma often not solved until it's too late. The ideal glueline is as thin as possible, but without starved spots. Thicker lines are generally weaker because they contain air bubbles or trapped solid particles, as well as internal stresses that develop as the adhesive shrinks during curing. Most glues, particularly PVAs, perform best if they're spread on both surfaces, and the surfaces then placed together and allowed to stand for about ten minutes before being clamped. This "closed time" gives the adhesive time to penetrate and coalesce before the clamps squeeze it out.

To bond successfully, glues require surprisingly little clamping pressure—10 psi is plenty, more will just squeeze out the glue, starving the joint. The most common clamping problem is an uneven glueline caused by poorly distributed pressure. Obviously, each job calls for its own setup, but a joint is clamped correctly when the glue squeezes out a bit just as the two parts mate, gap-free. Exert more pressure after that, an you risk starving the joint or racking the assembly. For edge-gluing, a good rule of thumb is to space clamps at intervals equal to twice the width of each board. So two four-inch boards should be clamped every eight inches, with generously dimensioned clamping blocks to spread the pressure and to protect the wood. Before actually gluing, dry-clamp your parts. If a joint won't close, fix what's wrong so you won't be tempted to draw it up later with crushing clamp pressure, introducing stresses that make failure probable.

George Mustoe followed this article with a second part for *Fine Woodworking*. He elaborates in tremendous depth on synthetics, how they solve some problems and pose new ones. The following, as with the first part, is reprinted with the permission of the magazine and Mustoe:

GLUES FOR WOODWORKING

Adhesives have changed enormously since the days when artisans prepared their own crude glues from meat scraps or milk curds. Twentieth-century chemistry has given us hundreds of new synthetic adhesives, some of which are of interest to the woodworker. Generally these adhesives—epoxies, cyanoacrylates, hot-melt glues and contact cements—are far more expensive than the hide glues, polyvinyl acetates and water-resistant glues, although they aren't necessarily more effective. In the small woodshop, cost alone limits use of most of these glues to special jobs.

Epoxy Resins

Although epoxies are among the more expensive adhesives, their physical properties—high strength, low shrinkage, transparency, insolubility, and ability to bond to a diverse array of materials—make them ideal for certain applications. The extreme strength of epoxy is seldom essential in joining wood to wood, but it makes it possible to bond wood to glass or to metal. Cured epoxy machines well, and its dimensional stability makes it an excellent choice for filling gaps and mending holes.

All epoxies are two-part systems: a resin, and a liquid catalyst or hardener. They harden by chemical reaction between the two components, not by solvent evaporation. The glue is activated by mixing the resin and hardener together, usually in equal proportions. Changing the resin/hardener proportion affects the properties of the cured epoxy, slightly increasing the hardener by up to 10 percent makes the bond more brittle. Using a larger proportion of either component weakens the bond.

Curing time for epoxies varies according to temperature. Epoxies generally require temperatures of 65 degrees F or higher to set, although special formulas have been developed for use at lower temperatures. Heating the joint to 100 degrees F to 150 degrees F speeds the setting rate, but also increases the health risk, the vapors from hot epoxy are very toxic. Viscosity drops dramatically at higher temperatures, causing the epoxy to flow out of the joint onto other areas of the work. Whatever the temperature, uncured epoxy is toxic, and repeated skin contact can provoke allergic reactions in some people. Acetone is recommended for cleaning up uncured epoxy, but denatured alcohol works just as well and is less flammable.

Epoxies come in a variety of different types. "Quick-set" or five-minute epoxies are convenient where you need a strong, fast-setting bond, but their brief pot life can be frustrating if your assembly takes longer than you expected. They also have less strength and water-resistance than conventional epoxy. Hand-moldable sticks of epoxy putty are easy to mix, and work well as a filler. Opaque "filled epoxies" contain suspended solids such as clay or powdered metal to provide increased strength, higher viscosity, or other desired properties. Filled epoxies have a putty-like consistency that makes them perfect for filling large voids or repairing surface dents, but they have a limited shelf life because the filler eventually settles out of suspension. Vigorous stirring will sometimes restore old stock to a usable condition. When clear epoxy resin gets old, it may become thick and granular (some preparations have recommended shelf lives of only 6 to 12 months), but warming the container to about 100 degrees F in a hot water bath will return the epoxy to its original state.

Many retail brands of clear epoxy are bought in bulk from the manfuacturer and repackaged into small containers. There isn't much difference between brands, and you can save money by purchasing epoxy in larger quantities. One ounce of epoxy in a tube costs about $2.25, but Sig Model Airplane Epoxy, an excellent transparent glue found in hobby stores, costs about $7.50 for 12 ounces. Though epoxy manufacturers won't often sell bulk quantities directly to the public, they'll usually provide technical assistance and lists of local distributors. Armstrong Products Company, PO Box 647, Warsaw, Indiana 46580, makes a clear all-purpose epoxy called A-271 resin, a "quick-set" resin called A-36, and several other types. Devcon Corporation, 30 Endicott Street, Danvers, Massachusetts 01923, also distributes epoxy in bulk. As another example of the money you can save by buying in quantity, Devcon 210 epoxy costs $2 an ounce in hardware stores, but about 62¢ an ounce in gallon lots.

Several epoxies have been marketed specifically for woodworkers by Industrial Formulators of Canada, Ltd. Their G-1 epoxy is a general-purpose clear resin; G-2 is recommended for oily and acidic woods such as teak and oak. Cold cure is meant for use a temperatures as low as 35 degrees F. Five Cure sets in 15 minutes or less at temperatures above 40 degrees F, and Sun Cure is a low-viscosity laminating resin. You can mail order these epoxies from Flounder Bay Boat Lumber, 3rd and "O" Ave., Anacortes, Washington 98221.

An extensive line of epoxies, additives, and dispensing pumps is sold by Gougeon Brothers, Inc., 706 Martin Street, Bay City, Michigan 48706, under the trademark West System. Developed for boatbuilding, the Gougeon system uses epoxies both as adhesives and as saturation coatings to prevent transfer of moisture and improve dimensional stability.

Polyester Resins

If you want to reinforce wood with fiberglass cloth, your best choice is epoxy resin, but because of epoxy's high cost, polyester resin is commonly used instead. It is also less toxic and much cheaper. Like epoxies, polyester resins are two-part systems: a low viscosity liquid which hardens when a small amount of catalyst is added. Although polyester resin performs well for reinforcing fiberglass, it lacks sufficient wetting ability to bind to wood fibers. It will adhere to wood only if the solidified resin can interlock with surface irregularities. You can get an adequate bond if you roughen the surface with a rasp or coarse sandpaper. Brush the catalyst-activated resin over the wood and allow it to soak in thoroughly. Before it begins to set, apply glass cloth and brush another coat of activated resin over the cloth. See Table 8-2.

Unlike epoxy, polyester resin shrinks considerably after curing, and it may remain slightly tacky long after solidification. To produce a smooth surface finish, polyester finishing resins sometimes contain emulsified wax which floats to the surface as the resin cures.

You may be able to save up to 50 percent by purchasing polyester resin and fiberglass cloth from a local business that uses them. Boatyards are usually willing to sell a gallon or two of resin from their 55-gallon drums at minimal markup. These industrial grade resins are sometimes slightly red or purple in color, compared to the water-clear retail resins. You can color clear polyester resin by adding pigments specially made for this purpose.

Cyanoacrylate Glues

No adhesive has received more attention in the last few years than cyanoacrylate, commonly knows as "super glue." Rumors abound, but cyanoacrylate is not made from barnacles. Nor is it new—the first cyanoacrylate adhesive, Eastman 910, was discovered by accident during a test of the high-refracting properties of a new organic compound when a drop was placed between glass prisms and they stuck fast. The glue was patented and first marketed in 1958, and industry has been using it ever since.

Cyanoacrylate will bond most plastics and rubber, and is good for gluing rubber to wood or to metal. Higher viscosity formulas are sold for use on wood and other porous materials. Elmer's Wonder Bond Plus and Krazy Glue for Wood and Leather are two brands available in small retail packages. As most users have discovered, cyanoacrylate also has a remarkable ability to bond skin. Glue distributors now sell solvents to dissolve unwanted bonds, although acetone and nail polish remover are somewhat helpful in this purpose.

Cyanoacrylate provides a very water-resistant, but not completely waterproof, bond. Prolonged immersion in water eventually weakens the joint. This adhesive will also resist most organic solvents.

Cyanoacrylate glue is most useful in mod-

Table 8-2. Specialty Woodworking Adhesives.

Adhesive	Application characteristics	Properties after curing	Recommended uses
Epoxy Armstrong A-271, Cold Cure, Devcon 210, Five Cure, Industrial Formulators' G-1 and G-2, Sig Model Airplane Epoxy, Sun Cure, West System	Moderate viscosity, decreasing greatly at warm temperatures, min. curing temp. 65°F**, excellent gap-filling ability, two-part system, liquid and vapors are toxic, may cause skin irritation	Highly transparent, waterproof, bonds to many materials, sands and machines well	Marine and outdoor use, excellent for bonding nonporous materials
Cyanoacrylate Devcon Zip Grip 10, Duro Super Glue, Eastman 910, Elmer's Wonder Bond, Hot Stuff, Krazy Glue, Scotchweld CA-3	Very fast bonding***, very low viscosity, min. curing temp. 60°F, poor gap-filling ability, odor may be irritating, bonding may be inhibited by oil or acidic residues, excess is very difficult to clean up	Highly transparent, very water-resistant	Small repairs, modelmaking, bonding nonporous materials
Hot-melt stick Bostik Thermogrip, Swingline Fix Stix	Sets almost instantly, high viscosity, excellent gap-filling ability, does not penetrate well, difficult to apply over large areas, nontoxic, glue gun needed	Neutral opaque color, moderate strength, remains slightly flexible, cannot be sanded, softens when heated	Furniture repairs, small projects such as toys, construction of jigs and patterns
Hot-melt sheet	Instant bonding, applied with hot iron, poor gap-filling ability		Veneering
Contact cement, solvent-based Weldwood Contact Cement, Wilhold Contact Cement	Instant bonding, min. curing temp. 70°F, poor gap-filling ability, highly toxic, very flammable, difficult to clean up	Low strength, creeps under load	Bonding plastic laminates to plywood, not recommended for veneering, though it is widely used for this purpose
Contact cement, water-based (latex)* Elmer's Cabinet Maker's Contact Cement, Weldwood Acrylic Latex Contact Cement	Instant bonding, min. curing temp. 65°F, poor gap-filling ability, low toxicity, easy to clean up, "open time" must be carefully monitored according to label directions	Very low strength, water-resistant, creeps under load	Recommended when limited ventilation conditions prevent using solvent-type cement

* Water-based adhesive may cause warping of veneer or thin panels.

** May be rapidly heat-cured at 100°F to 150°F.

*** Surface activator speeds set.

elmaking, musical instrument building, and other small-scale applications. Its main advantage is its extremely rapid set—3M's Scotchweld CA-3, for example, sets in about 30 seconds. You can reduce this setting time to as little as one second by brushing on 3M's Scotchweld Surface Activator for Cyanoacrylate Adhesives before you apply the glue. Moisture will also speed curing, but on wet surfaces the glue will leave chalky stains. I sometimes use the moisture of my breath to humidify small parts before gluing. The strength of the cyanoacrylate bond continues to increase slowly during the first 48 hours.

Shelf life of most cyanoacrylates is about 6 to 12 months—the glue thickens as it gets old—but storing the adhesive in the refrigerator will prolong its useful life. Since moisture speeds the setting time, however, allow the container to warm up to room temperature before you open it, or condensation in the bottle will offset the advantage of refrigeration.

Early cyanoacrylates did not work well on wood or other porous materials because the glue's viscosity is extremely low. This same property, though, allows the adhesive to penetrate hairline cracks. Cyanoacrylate dripped into a ragged break will reinforce the fracture, and it can strengthen joints already glued with another adhesive that have loosened slightly if it is dripped along the existing glueline. You can repair a large gap by packing it with baking soda, then dripping glue on, which turns the powder into a hard, white filler. To fill a small crack or hole, put a few drops of glue in the crack, then sand immediately with wet or dry sandpaper before the glue sets. The wood dust mixes with the glue to form a hard filler that matches the color of the wood. This works best on dark woods; on light colored woods the patch will be slightly darker than the surrounding wood. On

thin stock, it's a good idea to put a piece of masking tape on the back of the crack, to keep the glue from sticking to the bench. Cyanoacrylates do not set as quickly if there is acid present in the joint. If you're gluing an acidic wood such as oak, you'll get better results if you prepare the surfaces by brushing on a surface activator. These activators are mildly basic and neutralize the acid.

Cyanoacrylate is the most expensive adhesive generally available. Like epoxy, most brands are purchased in bulk and repackaged for retail sale, but you can save by buying in larger quantities. C.F. Martin Co. (of guitar fame), 510 Sycamore Street, Nazareth, Pennsylvania 18064, sells 3M Scotchweld CA-3 in one ounce containers for $12, Scotchweld Surface Activator for $8, or a kit consisting of two ounces of CA-3 and one container of surface activator for $25. Although there isn't much difference in glue quality between brands, there is a difference between the containers the glue is packaged in. Some styles clog before the adhesive is gone, when glue solidifies near the tip. Polyethylene dropper bottles are less likely to clog than metal squeeze tubes or rigid plastic containers. It helps to clear the nozzle by squeezing a little air out of the upright container before closing it.

Hot-Melt Glues

Synthetic hot-melt glues are easy to apply, and they set up quickly. Most are made of polyamide resins which melt at around 400 degrees F. Hot-melts are widely used in industry, where their quick set is an advantage on assembly lines. Their good gap-filling properties make them ideal for repairing worn, sloppy joints in old furniture. Hot-melts form thicker gluelines than most other adhesives, and have relatively low strength and poor penetrating ability. They're good for temporary jigs or tack-on fastenings, where extreme strength is not required. They're also well suited for joints that may need to be disassembled, but the heat necessary to break the glue bond will also damage the surrounding finish. Hot-melts develop 90 percent of their final bond strength within 60 seconds. The glue remains somewhat flexible and does not sand well. When the glue has cooled, the excess can be removed with a sharp blade.

Hot-melt sticks are sold for use in an electric glue gun. Manufacturers make several grades that cool at different rates. Those sold in retail stores allow you only about ten seconds to assemble parts, but you can increase open time slightly by preheating the parts. Hot-melts are also sold in thin sheets for veneering (available from Woodcraft Supply). You can use an ordinary household iron to provide heat, then weight or clamp the veneer until the glue has cooled.

Contact Cements

Contact cements are rubber-based (usually neoprene) liquids that dry by solvent evaporation. They are used most often to bond high-pressure plastic laminates, such as Formica, to plywood or particle board, without the need for clamps or prolonged pressure. Contact cements are sometimes used to attach veneers, but the glue bond can fail in spots because of seasonal moisture changes in the veneer, causing bumps in the veneered surface or separation at the edges.

There are two types of contact cements, solvent-based and water-based. Solvent-based cements, most of which are extremely flammable, dry in about five to ten minutes. The nonflammable solvent-based cements are made with chlorinated hydrocarbons, and their vapors are toxic. These vapors are not trapped by an organic-vapor respirator, so you should use adequate ventilation with this and any solvent-based adhesive. Water-based contact cements are nontoxic and nonflammable, but they take longer to dry—about 20 minutes to an hour before parts can be assembled. The uncured adhesive is water-soluble, so you can clean your tools in water if the glue hasn't dried. Water-based contact cement provides better coverage than the solvent-based type, but it should not be used on metallic surfaces.

Contact cements are heat-resistant and water-resistant, although adhesive strength is low and the pliable glue film is likely to creep under load. Both types can be applied by brush, roller, or spray. Apply adhesive to both surfaces to be mated and assemble when dry, but be sure that the parts are properly aligned. Adjustment is impossible once the two surfaces contact. Go over the glued surface with a roller to ensure an even bond.

"Construction adhesive" is a thick mastic used by carpenters to fasten flooring or wall paneling. As yet, it has not been widely used in other areas of woodworking.

A few other adhesives that may have limited use in woodworking are the acrylic cements, most commonly encountered as pressure-sensitive contact adhesives. Liquid acrylics are also used in some linoleum cements and other mastics. "Anaerobic adhesives" remain liquid when exposed to air and solidify when deprived of oxygen. The Loctite Corporation markets a variety of anaerobic adhesives which are widely used for securing nuts, bolts, and threaded studs. Because of its high porosity, however, wood contains too much oxygen to allow anaerobic adhesives to set.

WHY DIDN'T IT HOLD?

Why do glued joints fail? The reason is seldom because the glue is not adequately strong. Assuming the mating surfaces fit reasonably well, bond failure is most likely to be the result of one of three factors:

1. Choice of inappropriate adhesive
2. Inadequate surface preparation
3. Poor glue application or clamping technique

Choosing the Proper Adhesive

The choice of the proper adhesive is generally dictated by two very different sets of criteria. The first are controlled by conditions at the time of application: temperature, humidity, desired setting rate, and the method of application, as well as the composition of the materials being joined. The second criteria involve performance characteristics after curing, such as the ability to withstand high or low temperatures, water resistance, tendency to creep under load, machinability, sandability, and color of the glue line. Fortunately, adhesives are available which meet almost every combination of requirements.

Surface preparation is somewhat more complicated with wood than most other materials. Adhesives typically bond primarily to the wood surface, and the total depth of glue penetration is of less importance. Fibers at the surface should be intact, not torn or partially detached as the result of cutting with dull tools. Fine sanding is not desirable, as dust fills pores and prevents glue from contacting the solid underlying wood. Epoxy and polyester resins may bond better to wood that has been roughened with a rasp or coarse sandpaper, but in other cases bond strength may suffer because of damage to the surface wood fibers. Woodworkers commonly observe that joints fail not by parting of the adhesive film, but because of separation of the wood adjacent to the glue line. This phenomenon results because of damage to the fibers prior to gluing. Microscopic structural damage always occurs when a piece of wood is dimensioned,

but the best joints result when the surface is planed or scraped smooth. Fuzzy, abraded surfaces left by saw teeth or coarse abrasives may soak up more glue, but the final bond is not as strong.

Surface Preparation

Before gluing, all dust should be removed with a damp rag. Oily or pitchy woods can be wiped down with mineral spirits to remove surface residues, and the wood must be allowed to dry before glue is applied. Mating surfaces should fit well, but joints should contain enough space to permit a glue film to develop; hammer-tight tenons or dowels are not recommended. Sloppy fits are a problem during repair of worn furniture, and the best cure is to use a glue with wood gap-filling properties, such as epoxy or hot-melt adhesive. A badly cut joint can sometimes be salvaged by removing enough stock to allow insertion of a veneer shim, avoiding the need for a massive glue layer. Open-grain woods and end grain surfaces may soak up glue, leaving a weakened or starved joint. The problem of porosity can be eliminated by applying glue liberally and waiting several minutes for the wood to become saturated before applying clamping pressure.

Starved joints can also result from excessive clamping pressure, but the most common clamping problem is poorly distributed pressure, producing an uneven glue line. The ideal glue joint is one which is as thin as possible, but without starved areas. Thicker glue films generally have less strength, and they are more likely to contain air bubbles or trapped solid particles, as well as having internal stresses that develop as the adhesive shrinks during curing. Thick glue lines have greater resistance to cracking or creeping under stress, and they represent most efficient use of the adhesive.

Water-Based Adhesives

Water-based adhesives may cause temporary internal stresses to develop as moisture invades the surrounding wood. This swelling can warp thin panels or veneer. Ordinarily, these distortions would go away when the wood returned to its normal moisture content, but in the meantime curing of the glue may lock in irregularities. For the same reason, lumber should not be glued during conditions of unusually high or low atmospheric humidity. Similarly, heat-curing of epoxy or plastic resin glue can cause distortion as moisture is given off, and delicate items such as musical instruments should never be heat-cured. Thick wood is a very good insulator, a physical property that must be considered during heat curing.

During World War II, one and two-inch thick lumber was laminated into keels and ribs for navy vessels. These timbers were typically 12 inches wide and 24 inches deep. When they were placed in a 180 degree curing oven, measurements showed inner gluelines did not reach this temperature for a day or longer. Yet veneers glued with the same adhesive would bond in a few minutes. Heat curing can also be accomplished by high frequency induction. This process is similar in principle to a microwave oven; heat is generated only within the glue layer while the surrounding wood remains cool. This process allows rapid gluing, but the expense of the equipment generally limits the technique to industrial processes.

Chapter 9

Craft and Hobby Glues

Most hobbies and crafts require glues and adhesives. Some of the most beautiful objects are put together with special preparations. The adhesives must possess special characteristics and properties.

WORKING WITH PLASTIC

Polystyrene cement is the most popular glue used in the construction of plastic kits or objects. It is used so much it is often called plastic cement. Cements are composed of plastic dissolved in a solvent. There are different types since some are made specifically to bond plastic to plastic. One of the most basic formulas is a solution of polystyrene in chlorobenzene or acetone. It comes in a transparent, viscous fluid in tubes, commonly named tube cement, and in a liquid, naturally called liquid cement.

The solvents are powerful, and model makers must use them with great care. The tube cement tends to be stronger than the clear liquid. If too much is applied, and it squeezes from a joint onto the surface of the kit or object, it may not only remove the detail in the area, it may even eat a hole in it! Getting some excess even on the interior or back of the kit or piece may distort the molding, even if it has been painted after the glue work was done. The dissolving may continue for a long time. Tube cement is best for bonding structural components or rather loose-fitting joints. The more volatile liquid cement should be used if there is any danger of damaging the piece. Never use the tube or liquid cement on polystyrene foam because it will dissolve it. White glue is best for this material.

When joining two halves of a model or craft piece, use a tiny drop of tube cement on each of the two locating pins. Rubber bands or tape serve as the best clamps to secure the pieces while the gluing is being done. After the cement is applied right up to the tape or bands, they should be removed, and using a pointed brush cover the area where there still needs to be glue. With structural components, tube cement should be applied with precision. After the pieces are joined, the joint should be sealed with liquid cement.

Attaching smaller pieces involves tacking them in place with liquid cement first. A cement line should be run around the item to ensure it attaches. In some ways, it is better to apply the solvent to each surface, softening them before pressing them together. The highly volatile solvents make for fast-setting cements. Using compatible cements, many plastics can be bonded, but not all of them will readily work on nylon and polyethylene. In essence, the solvents soften the surfaces to be mated, fusing the parts as the solvent evaporates.

BALSA WOOD

Mechanical adhesion works well with balsa wood and other very porous materials. Technically, this is not a cementing process. The two or more surfaces form an interlocking bond. Synthetic resins are added to balsa cements to give them strength and to make them better for filling gaps. Balsa cement must always be put on both surfaces. The secret to getting end-grain porous pieces to bond well is double-gluing. Since the ends soak up so much glue, they can starve the glue line, not leaving enough to get a strong bond. Apply the cement to each surface and press them together. Then, immediately pull them apart, letting the cement get fairly dry without setting. Put another layer of cement on each surface and attach them. Balsa cements are not waterproof.

Balsa wood is used extensively in modelmaking and in many arts and crafts. The cements are better suited for small areas. Some people prefer to use white glue for working with larger surfaces because it is very effective.

ACRYLIC CEMENTS

Very thin acrylic cements work best with modelmaking and many crafts. They contain acrylics dissolved in dichloromethane. It is wise to check out the compatibility of the cement with clear acrylics, since these can fog or discolor. Care must be taken to ensure that no air bubbles get trapped in the gluing. Some acrylics are very expensive, so testing is important.

Before using acrylic cement, the joint should be cleaned with a solvent such as cigarette lighter fluid and then washed with clean water. Putting masking tape around the area of the joint will prevent any cement from getting on the surfaces where it should not be. Putting a strip of tape on one side of the two pieces to be joined will help align them while the cementing is done. Using the tape as a hinge, pull the joint apart, and apply the cement to both pieces. Open time should be no more than 30 seconds before mating the joint. It should set for at least three hours, and the joint will not really be cured for some 20 days.

WHITE GLUE

This general woodworking glue will mate different types of wood as well as bonding many other materials. Since the objects are light in weight, the glue will not allow them to creep apart since there is little stress. Both surfaces must be coated with white glue, and clamping is standard for at least 24 hours. Keep a damp cloth handy to remove any excess white glue since it is difficult to get off after it hardens. Joints that fit snugly and really hang on to each other may not need clamping. Moisture will tear apart white glue connections because it softens in water. In some rare cases, white glue may need to be thinned with water.

UREA FORMALDEHYDE

UF glues are synthetic resins that are ideal for many woodworking tasks. UF is waterproof and great for boat modeling even though the setting time varies. It should be applied to both surfaces, leaving them apart for a while before mating them. It is best to clamp the joints overnight. When it is warm where the work is being done, the glue will set in two to three hours. It should never be used when the temperature gets below 50 degrees F, since the coldness interferes with the chemical reaction that makes the adhesion.

CNA GLUE

Cyanoacrylate glues have many uses in hobbies and crafts. Their instant bonding of 10 to 15 seconds is one of their main assets. Although they are costly, they usually require only one drop so in proportion they are really no more expensive to use. The clear liquid must be used in very tiny amounts, since too much will make a weaker instead of a stronger bond. They are too thin to have any gap-filling properties, so joints must fit closely. It is possible to bond the joint first with the CNA then add a gap-filling glue around the area to close it off. An epoxy works well as a filler.

For most jobs, the glue can be applied directly from the little containers, but smaller jobs may require the liquid to be put on with a toothpick. The thinness can present a problem on porous surfaces. Wetting the surfaces to be joined slightly will prevent a problem on porous surfaces. Wetting the surfaces to be joined slightly will prevent the glue from penetrating too much. When a joint refuses to stick, take it apart and let it dry for some ten minutes. When you get ready to glue it again, use less of the liquid. Contact with the skin should be carefully avoided—the stuff will bond your fingers quickly. Always keep the container pointed away from you when opening or closing it. Certainly, take great precaution around the eyes.

Cracks in porous materials can be made very strong by soaking the cracked areas with CNA. It can also be used to give strength to pieces that are not broken or cracked but simply need some reinforcement.

RUBBER CEMENT

This is a big category of adhesives that use natural or synthetic rubber in solutions. They are universally accepted as excellent when working with paper and cardboard. They will bond flexible materials to each other, or bond flexible things to rigid surfaces. Technically, these are rubber-based adhesives while the rubber cements are rubber solutions with resin additives to give them strength and the ability to fill gaps. Those based in water are usually labeled as latex adhesives. Care must be exercised with the spirit-based rubber cements since they are highly flammable. They must never be near fire or very hot objects.

Many of the rubber adhesives are contact glues, so they must be applied to both surfaces and left to dry until they are just a little tacky before joining. They bond immediately, and they get stronger as the solvent evaporates. It takes the usual 24 hours to get the proper cure. Residue is easy to remove by just rubbing it—it literally balls up. One thing to avoid with these cements is photographs—it will usually discolor them.

Lapping one piece of paper or cardboard over another makes a strong joint when gluing. With cardboard or balsa wood projects, gluing cardboard or balsa strips to give support at stress points is recommended. Backing paper with cardboard is easy when the rubber cement is spread evenly on both surfaces. After they dry, put a sheet of wax paper over the cardboard, leaving it exposed at one end. Then rub the paper down onto the exposed section. Ease the wax sheet out a bit at a time until the paper is fitting well. The fact that rubber cements don't leave any wrinkles in paper is one of their greatest properties.

There are some specialty glues in making models and crafts. Acrylobutyl styrene, ABS, solvent will bond many plastics, including polystyrene and acrylic. There is also a tissue paste that is used for applying thin, delicate coverings to model planes and other things. This dextrin adhesive or dope may be used for fine covering jobs, including nylon and silk model aircraft coverings. Tissue paste or tissue cement will work as well as fine dope.

FABRIC GLUE

Making, detailing, and clothing dolls is a popular hobby or craft, and the right glues can make a "million dollar baby" out of some modest materials and creativity and skill. Doll costumes can be made by sewing, gluing, or a combination of techniques. Handstitching a tiny garment takes great skill and much time since the stitches must be practically invisible. Stitches that show are too large to be in proportion with the whole doll.

An alternative to the many hours of minute sewing is fabric glue. This is the quickest and a very acceptable way to join two pieces of cloth. Not only is gluing fast, it is very durable. The bond will last as long as the fabric—probably longer. Miniature dolls require some sewing in addition to the gluing. It is easier to gather skirts or cuffs with a needle, but accessories can be readily attached with glue. The fabric must be compatible with the glue. Most cottons and fabrics with a high percentage of cotton work extremely well with glue. Most fabric glues will not work with many man-made fibers. Materials with a nap, such as suede or velvet, are exceptions.

There are several types of fabric glue, and some people do a most beautiful job by gluing the garments directly on the dolls. Larger fabric stores usually have a selection of glues.

Chapter 10

Sealants Save Energy

Sealants are usually synonymous with adhesives—you can't seal something unless it sticks. Keeping out weather and water is basic to man's shelter, and some sealants are needed in all homes to varying degrees.

Whenever two different kinds of building materials meet, there is usually a need for a caulking. Vent pipes and chimneys usually require flashing attention from time to time on most houses. Although flashing joints are sealed with roofing cement when they are built, years of weathering wear away the joints. These leaks can be stopped by repairing the flashing. Sealing the worn areas with liberal applications of roof cement will usually wear well for several more years. The cement should go right on up the pipe or chimney a short way to get a good bond.

A caulking or roof cement should be used to fasten the flashing to the surface, making a total seal to prevent any water from entering.

It is best to replace the flashing any time there has been leaking. Most patching turns out to be temporary.

TYPE OF CAULKING

Oil based caulking is the most common and generally least expensive. It bonds to wood, masonry, and metal. This factor makes it popular.

Latex-based or water-thinned caulking dries quickly and sticks to paint pretty well. Some of the poorest caulkings never seem to dry any, and they should.

Butyl rubber caulking is good for flashing jobs since it bonds metal to masonry well. It is also long-lasting when of a high grade.

Polyvinyl acetate caulking will stick to all surfaces. including painted ones. It, too, is very durable.

Silicon caulking is the most durable type, and it bonds well to all surfaces with the exception of painted ones. Most caulking compounds come in white, gray, or black; some of them do not take paint very well, while others do. If the caulking is going to be very visible, this should be taken into consideration.

Filling large cracks can take a tremendous amount of caulking. Any big cracks should first be filled with oakum, packing the rope-like material into the crevice. The filler is much cheaper than caulking. When a large amount of caulking may be needed, such as for a remodeling or major repairs or in new home construction, consider buying caulking in bulk. Some firms charge a considerable amount more for their caulking when it is packaged in the cartridges. Some compounds can be bought in coils like rope. The caulking is just unwound and applied.

WHERE TO USE CAULKING

Windows and doors allow more entrance and exit of energy than anything else in a house. Even the most energy-efficient windows will not keep air from coming in around the sills; only caulking will stop the air. Caulking is needed where the sill meets the window frame. Caulking the window pane to the frame is one of the most common repair jobs. Windows composed of small panes need a lot of caulking work after so many years.

Door frames funnel in air when they have not been caulked. Two pieces of wood meeting in a door frame invariably do not fit airtight. Porous weatherstripping usually will not do the job as well as caulking. Old caulking should always be removed before applying new. After scraping off the old caulking with a chisel, scraper, or screwdriver blade, clean the area with a solvent. Let dry before applying.

Caulking should seal where dormers or any other protrusions come off a house. Any break in the solid structure should be sealed.

Every corner needs sealing, especially those formed by siding. The siding beneath windowsills needs caulking. Gable moldings and the undersides of eaves need sealing.

Learning to caulk well takes some careful practice. Getting a good bead is the objective, and it must lap both surfaces. The gun should be held at a 45 degree angle and slanted towards the direction of movement. There are some packages of caulk that come in their own dispenser; these are similar to giant tubes of toothpaste. Wide beads are necessary in many applications.

Hundreds of tons of caulking are used in fixing bathroom problems. The most common bathroom plague is the inevitable crack that seems to appear between the tub and wall. The tub is filled with water daily and then drained out. This constant weight change eventually pulls the tub away from the wall. A waterproof, flexible compound must be used around the tub rim. If the crack is rather wide, there are rubber fillers that can be caulked into the area.

EXTERIOR WATERPROOFING

Water is always trying to get into houses and buildings. A bad flashing will quickly cause water damage to a structure. Wood shingles and shakes are particularly prone to water. They should be sealed periodically. We prefer Thompson's Water Seal for the finest exterior wood protection. It is extremely easy to apply with a brush, roller, or spray.

Thompson's Water Seal dries clear and forms a protective shield. Wood windowsills and frames need protection from moisture. There are literally millions of rotted sills out there in the nation now. A good sealer would have saved them. Wood shutters and even canvas awnings can be protected with the Water Seal. Wood railings, wood decks, and wood flower planters must be sealed from the weather or suffer deterioration. Outside furniture will quickly weather if it is not protected, and nothing is more outdoors than a wood garage door.

The Water Seal works just as well at protecting brick, concrete, and terra cotta. Brick walkways, concrete steps, and concrete driveways need to be covered with the Water Seal. Stucco foundations quickly show if they have not been sealed. Chips, cracks, and discoloration start to appear. Water can start silently destroying a house or building in some 15 places as has just been cited. Thompson's Water Seal may be found in many paint, hardware, and home and building centers.

Caulking and sealing are among the major energy-saving investments. Good caulking can perform a variety of duties in addition to the obvious ones. Dryer exhaust vents and air conditioners should be sealed to ensure energy efficiency. Water, gas, or oil line entries into a structure should be caulked. Skylights (which seem to have a high potential for leaking anyway) need to be perfectly sealed and checked periodically.

The joint between a foundation and driveway should be filled to prevent water from damaging both of them. Patio covers should have their seams sealed from the weather, and metal storage buildings will do their job well when they are protected with caulking. Be sure you get a compound that will not crack, chip, or peel.

Sealers keep out things besides water, too. Ozone, the sun's ultraviolet rays, and any of thousands of pollutants can be kept out of a structure to some degree by adhesive and sealants.

Mineral spirits will clean up most caulking, and it should be applied only in well-ventilated areas. Caution is directed when caulking large areas in an enclosed space. Some caulking does not allow very much working time. Some directions note that the compound should be tooled as it is going to be within five minutes after it is applied.

There are special sealants for marine use, depending on whether the product is going to be used above or below the water. Boat hatches and windows need periodic attention as to the condition of the caulking. There are places inside the cabin where sealers should be used to maintain the best conditions. Boathouses tend to suffer water damage quickly if they are not well-sealed.

Homes located on or near water need their wood protected from the humidity. Caulking and sealants are as necessary for protection as the windows and doors.

Sealing or caulking is the primary job before painting a structure. After all the scraping has been done on exterior walls, the caulking must be done. Paint alone will certainly not seal a building. The caulking must be put everyplace where there might be water penetration. Just prior to painting is definitely the time to check the caulking and flashing and inspect every structural joint for its sealant needs. A putty knife is one of the best tools for smoothing caulking, but the tool must match the application.

There are many specialty sealants and caulkings for use in many fields of industry. Food-handling and processing plants must use compounds that will present no hazards to health. For various building codes, the caulking must meet particular standards. Caulking is a good warm-weather job since it does not work well when it gets into the 40-degree range. Outdoor receptacles need caulking at the juncture between the cover plate and the wall surface. Not only does this ensure that the outlet is waterproof, but that it also prevents any air from entering the structure at that point. Even lights under eaves should be protected with the compound. Yard lighting should also be sealed well.

Any points where an antenna is mounted on a building should be caulked. Caulking is used also where some connections are made on the satellite television dishes. Moisture can be the cause of some great reception problems.

Appendix
Adhesives Applications Chart

Although various materials are dealt with in detail throughout the book, the following chart outlines various materials and the glues and adhesives that work best with them.

BONDING MATERIALS

	Balsa Wood	Soft/Hard Woods	Plywood	Polystyrene	ABS	Acrylic/Plexiglas	Cardboard	Paper	Fabric	Polystyrene Foam	Poly Foam	Metals	Cork	Glass	Rubber	Fiberglass
Fiberglass	JL	JL	JL	N	N	N	JN	H	HL	EI	JL	JL	HJ	JL	CH	JL
Rubber	H	H	H	I	GJ	GJ	GJ		H	I	HN	GJ	HJ	HJ	HJ	
Glass	J	J	J	GJ	GJ	GJ	GJ	EI	H	EI	HJ	GJ	HJ	L		
Cork	HJ	HJ	HJ	J	J	GJ	GJ	EI	HI	EI	EI	HJ	EHL			
Metals	J	J	J	GJ	GJ	GJ	GJ	EI	H	EI	HJ	GJ				
Polyurethane Foam	EI	EI	EI	EIJ	EIJ	EIJ	EI	E	EI	EI	EHIJ					
Polystyrene Foam	E	E	E	EI	EI	EI	E	EI	EI	EI						
Fabric	HI	HI	HI	H	H	H	HI	EI								
Paper	E	E	E	E	E	E	DE	DE								
Cardboard	DE	DE	DE	EGJ	EG	EG	DE									
Acrylics/Plexiglas	GJ	GJ	GJ	BGJ	BGJ	C										
ABS	GJ	GJ	GJ	BGJ	BGJ											
Polystyrene	GJ	GJ	GJ	AGJ												
Plywood	EF	EF	EF													
Soft/Hard Woods	EF	EF														
Balsa Wood	DE															

KEY TO TABLE

A Polystyrene cement
B ABS cement
C Acrylic cement
D Balsa cement
E White glue
F UF glue/Synthetic Resin glue
G Cyanoacrylate glue
H Rubber base contact glue
I Latex glue
J Epoxy glue
K Clear dope
L Polyester resin

Index

OTHER POPULAR TAB BOOKS OF INTEREST